Male Germline Chromatin

Male Germline Chromatin

Editors

Darren Griffin
Peter Ellis

MDPI • Basel • Beijing • Wuhan • Barcelona • Belgrade • Manchester • Tokyo • Cluj • Tianjin

Editors
Darren Griffin
University of Kent
UK

Peter Ellis
University of Kent
UK

Editorial Office
MDPI
St. Alban-Anlage 66
4052 Basel, Switzerland

This is a reprint of articles from the Special Issue published online in the open access journal *Genes* (ISSN 2073-4425) (available at: https://www.mdpi.com/journal/genes/special_issues/Male-Germline-Chromatin).

For citation purposes, cite each article independently as indicated on the article page online and as indicated below:

LastName, A.A.; LastName, B.B.; LastName, C.C. Article Title. *Journal Name* **Year**, *Article Number*, Page Range.

ISBN 978-3-03936-854-9 (Hbk)
ISBN 978-3-03936-855-6 (PDF)

Cover image courtesy of Ben Skinner and Peter Ellis.

Contents

About the Editors

Darren Griffin received his Bachelor of Science and Doctor of Science degrees from the University of Manchester and University College London, respectively. After postdoctoral stints at Case Western Reserve University and the University of Cambridge, he landed his first academic post at Brunel University before settling at the University of Kent, where he's been for the last 15+ years. He has worked under the mentorship of Professors Joy Delhanty, Christine Harrison, Terry Hassold, Alan Handyside, and Malcolm Ferguson-Smith. He is President of the International Chromosome and Genome Society, a Fellow of the Royal College of Pathologists, the Royal Society of Biology, and the Royal Society of Arts, Manufacture, and Commerce. He sits on the faculty of CoGen (controversies in genetics) and has previously sat on the board of the Preimplantation Genetic Diagnosis International Society (PGDIS), organizing its annual meeting in 2014. Darren is a world leader in cytogenetics. He performed the first successful cytogenetic preimplantation genetic diagnosis (sexing of IVF embryos) and, more recently, played a significant role in the development of Karyomapping, a universal test for genetic disease in IVF, an approach he now applies to cattle. In his 30+ years of scientific research, he has co-authored over 200 scientific publications, mainly on the cytogenetics of reproduction and evolution, most recently providing insight into the karyotypes of dinosaurs. He is a prolific science communicator, making every effort to make scientific research publicly accessible (both his own and others) and is an enthusiastic proponent for the benefits of interdisciplinary research endeavor. He has supervised over 35 PhD students to completion and his work appears consistently in national and international news. He currently runs a vibrant research lab of about 20 people (including a program of externally supervised students) and maintains commercial interests in the outcomes of research findings, liaising with companies in the agricultural sector in the area of fertility screening. Darren is a member of the Centre for Interdisciplinary Studies of Reproduction (CISoR). He also regularly coordinates the International Chromosome Conferences and the Pig Breeders' Round Table.

Peter Ellis is a Lecturer in Molecular Genetics and Reproduction at the University of Kent. Key findings from his works include the identification of novel genes on the mouse Y chromosome that affect sperm head shape and fertility; the discovery of a genomic conflict or arms race between the X and Y chromosomes in mice as they compete to influence offspring sex ratio, which in turn has dramatically affected the structural and functional content of both chromosomes; and the identification of mechanisms regulating meiotic and post-meiotic transcriptional silencing of the sex chromosomes. His laboratory investigates the molecular biology of reproduction, the conflicting roles played by sex-linked genes in regulating this process, and the relationship between DNA damage repair mechanisms and the checkpoints governing meiotic progression.

Editorial

Form from Function, Order from Chaos in Male Germline Chromatin

Peter J. I. Ellis and Darren K. Griffin *

School of Biosciences and Centre for Interdisciplinary Studies of Reproduction, University of Kent, Giles Lane, Canterbury CT2 7NJ, UK; P.J.I.Ellis@kent.ac.uk
* Correspondence: d.k.griffin@kent.ac.uk

Received: 28 January 2020; Accepted: 9 February 2020; Published: 18 February 2020

Abstract: Spermatogenesis requires radical restructuring of germline chromatin at multiple stages, involving co-ordinated waves of DNA methylation and demethylation, histone modification, replacement and removal occurring before, during and after meiosis. This Special Issue has drawn together papers addressing many aspects of chromatin organization and dynamics in the male germ line, in humans and in model organisms. Two major themes emerge from these studies: the first is the functional significance of nuclear organisation in the developing germline; the second is the interplay between sperm chromatin structure and susceptibility to DNA damage and mutation. The consequences of these aspects for fertility, both in humans and other animals, is a major health and social welfare issue and this is reflected in these nine exciting manuscripts.

Keywords: spermatogenesis; chromatin; nuclear organisation; DNA oxidation; DNA fragmentation; epigenetic inheritance; histone retention; assisted reproduction; in vitro fertilisation

---

One of the most fundamental requirements in spermatogenesis is the need to develop male germ cells to undergo radical restructuring of their chromatin. Occurring at multiple stages before, during and after meiosis, it involves coordinated waves of DNA methylation and demethylation. It also involves histone modification, replacement and removal. In this Special Issue, we draw together novel studies and contemporary reviews addressing various aspects of chromatin organization and dynamics in the male germ line, and consider both humans and model organisms. Two major themes emerge from these exciting studies: the first being the functional significance of nuclear organization in the developing germline and the second is the interplay between sperm chromatin structure and DNA damage. The consequence of these aspects for fertility, both in humans and other animals, is a major health and social welfare issue.

Fernanda López-Moncada and colleagues address the question of whether chromosomal reorganization alters gene expression during meiotic prophase [1]. In particular, they show that Robertsonian fusions involving chromosomes bearing nucleolar organizing regions (NOR) perturb their normal organization and nucleolar functionality. In post-meiotic spermatids, Jonathan Riel and colleagues show that Sly deficiency is not the only reason for infertility in mice with deletions on their Y chromosome. Rather, it appears that some other Yq-encoded gene is likely to be required to allow Sly to bind to chromatin and to exert its normal regulatory functions [2].

Four studies examine chromosome organisation in mature sperm. First, Dimitris Ioannou and Helen Tempest show that, while chromosomes in human sperm do indeed to form hairpin loops, as predicted from studies in other species, their centromeres are not organized in the classic "chromocenter" arrangement seen in model species such as mice [3]. Second, Heather Fice and Bernard Robaire confirm that relative sperm telomere length does indeed decrease during ageing in rodents, but, crucially, only in inbred strains [4]. Moreover, the demonstration that relative telomere length changes as sperm pass through the epididymis is a novel one. Third, Ben Skinner and colleagues address the question of whether chromosome territory organization is conserved between species, demonstrating that mouse chromosomes have retained the same sub-nuclear "address" for over two million years of evolutionary history [5]. Finally, Alexandre Champroux and colleagues turn to the possible deleterious effect of oxidative damage on sperm DNA organization. The surprising finding is that territory organization is largely robust in response to this challenge, with the overall organization of the chromosome territories being maintained even in the face of oxidative DNA damage. However, this organization is then disrupted in response to the treatment, illustrated by the reducing agents, signifying that oxidative damage may perturb chromosome decondensation following fertilization [6].

The theme of DNA damage is covered extensively in our two review articles. While DNA damage is usually regarded as a pathological, abnormal process, Tiphanie Cavé and colleagues review the role of endogenous, naturally-occurring DNA strand breaks created during chromatin remodeling [7]. This is an emerging field with profound implications for our understanding of the processes generating structural variations and polymorphisms within the genome, and the male versus female bias of specific mutational signatures. In a similar vein, but with a more clinical focus, Jordi Ribas-Maynou and Jordi Benet take a look at the differential reproductive effects on male fertility of single and double strand sperm DNA damage, respectively [8]. By their account, single-strand DNA breaks are present as scattered break points throughout the genome, whereas double-strand DNA breaks are mainly localized and attached to the sperm nuclear matrix. Single strand breaks are related to oxidative stress and impede pregnancy rates, whereas double strand breaks may be related to a lack of meiotic DNA repair—or to genome reconfiguration by topoisomerases, as highlighted by Cavé and colleagues—and lead to increased miscarriage rates, low embryo quality and implantation failure during ICSI.

Finally, we are particularly proud of the use of novel methods for studying the interplay between chromatin structure and the susceptibility to DNA damage and mutation. Indeed, this Special Issue boasts three new methodological approaches with Sheryl Homa and colleagues comparing two means of measuring oxidative stress (concluding that both used in tandem are better than one in isolation) [9] and both the Skinner and Champroux papers taking novel approaches to quantify the localization of chromosome territories in asymmetrical nuclei [5,6].

Collectively, these papers serve to highlight the importance of understanding male germline chromatin organisation in order to appreciate how specific regions of the genome may well be exposed to different stressors, remodeled, and activated before or after others immediately following fertilization. This, in turn, has downstream effects on both male germline mutagenesis and for early embryonic development; with profound subsequent implications for understanding natural fertility and improving assisted reproduction techniques.

Taken together, this unique collection of studies will, we hope, serve as a benchmark for a deeper understanding of the fundamental mechanisms perpetuating our germline.

Acknowledgments: Ellis is funded by the BBSRC, grant number BB/N000463/1 and the Leverhulme Trust, grant number RPG-2019-194.

Conflicts of Interest: The authors declare no conflict of interest.

References

1. López-Moncada, F.; Tapia, D.; Zuñiga, N.; Ayarza, E.; López-Fenner, J.; Redi, C.A.; Berríos, S. Nucleolar expression and chromosomal associations in Robertsonian spermatocytes of *Mus musculus domesticus*. *Genes* **2019**, *10*, 120. [CrossRef]

2. Riel, J.M.; Yamauchi, Y.; Ruthig, V.A.; Malinta, Q.U.; Blanco, M.; Moretti, C.; Cocquet, J.; Ward, M.A. Rescue of *Sly* Expression Is Not Sufficient to Rescue Spermiogenic Phenotype of Mice with Deletions of Y Chromosome Long Arm. *Genes* **2019**, *10*, 133. [CrossRef]

3. Ioannou, D.; Tempest, H.G. Human Sperm Chromosomes: To Form Hairpin-Loops, Or Not to Form Hairpin-Loops, That Is the Question. *Genes* **2019**, *10*, 504. [CrossRef]

4. Fice, H.E.; Robaire, B. Telomere Dynamics Throughout Spermatogenesis. *Genes* **2019**, *10*, 525. [CrossRef] [PubMed]

5. Skinner, B.M.; Bacon, J.; Rathje, C.C.; Larson, E.L.; Kopania, E.E.K.; Good, J.M.; Affara, N.A.; Ellis, P.J.I. Automated Nuclear Cartography Reveals Conserved Sperm Chromosome Territory Localization across 2 Million Years of Mouse Evolution. *Genes* **2019**, *10*, 109. [CrossRef] [PubMed]

6. Champroux, A.; Damon-Soubeyrand, C.; Goubely, C.; Bravard, S.; Henry-Berger, J.; Guiton, R.; Saez, F.; Drevet, J.; Kocer, A. Nuclear Integrity but Not Topology of Mouse Sperm Chromosome is Affected by Oxidative DNA Damage. *Genes* **2018**, *9*, 501. [CrossRef] [PubMed]

7. Cavé, T.; Desmarais, R.; Lacombe-Burgoyne, C.; Boissonneault, G. Genetic Instability and Chromatin Remodeling in Spermatids. *Genes* **2019**, *10*, 40. [CrossRef] [PubMed]

8. Ribas-Maynou, J.; Benet, J. Single and Double Strand Sperm DNA Damage: Different Reproductive Effects on Male Fertility. *Genes* **2019**, *10*, 105. [CrossRef] [PubMed]

9. Homa, S.T.; Vassiliou, A.M.; Stone, J.; Killeen, A.P.; Dawkins, A.; Xie, J.; Gould, F.; Ramsay, J.W.A. A Comparison Between Two Assays for Measuring Seminal Oxidative Stress and their Relationship with Sperm DNA Fragmentation and Semen Parameters. *Genes* **2019**, *10*, 236. [CrossRef] [PubMed]

Article

Nuclear Integrity but Not Topology of Mouse Sperm Chromosome is Affected by Oxidative DNA Damage

Alexandre Champroux, Christelle Damon-Soubeyrand, Chantal Goubely, Stephanie Bravard, Joelle Henry-Berger, Rachel Guiton, Fabrice Saez, Joel Drevet * and Ayhan Kocer *

GReD "Genetics, Reproduction & Development" Laboratory, UMR CNRS 6293, INSERM U1103, Université Clermont Auvergne, 28 Place Henri Dunant, 63000 Clermont-Ferrand, France; alexandre.champroux@uca.fr (A.C.); christelle.soubeyrand-damon@uca.fr (C.D.-S.); Chantal.goubely@uca.fr (C.G.); Stephanie.bravard@uca.fr (S.B.); joelle.henry@uca.fr (J.H.-B.); rachel.guiton@uca.fr (R.G.); fabrice.saez@uca.fr (F.S.)
* Correspondence: joel.drevet@uca.fr (J.D.); ayhan.kocer@uca.fr (A.K.)

Received: 11 September 2018; Accepted: 15 October 2018; Published: 17 October 2018

Abstract: Recent studies have revealed a well-defined higher order of chromosome architecture, named chromosome territories, in the human sperm nuclei. The purpose of this work was, first, to investigate the topology of a selected number of chromosomes in murine sperm; second, to evaluate whether sperm DNA damage has any consequence on chromosome architecture. Using fluorescence in situ hybridization, confocal microscopy, and 3D-reconstruction approaches we demonstrate that chromosome positioning in the mouse sperm nucleus is not random. Some chromosomes tend to occupy preferentially discrete positions, while others, such as chromosome 2 in the mouse sperm nucleus are less defined. Using a mouse transgenic model ($Gpx5^{-/-}$) of sperm nuclear oxidation, we show that oxidative DNA damage does not disrupt chromosome organization. However, when looking at specific nuclear 3D-parameters, we observed that they were significantly affected in the transgenic sperm, compared to the wild-type. Mild reductive DNA challenge confirmed the fragility of the organization of the oxidized sperm nucleus, which may have unforeseen consequences during post-fertilization events. These data suggest that in addition to the sperm DNA fragmentation, which is already known to modify sperm nucleus organization, The more frequent and, to date, The less highly-regarded phenomenon of sperm DNA oxidation also affects sperm chromatin packaging.

Keywords: mouse sperm chromatin; chromosome organization; nuclear-3D-parameters

---

1. Introduction

The mammalian spermatozoon is a highly-differentiated cell produced by the testis during a long and complex process called spermatogenesis. Following successive steps that lead to the multiplication and the production of haploid germ cells through the meiotic program, spermatids undergo a long phase of cyto-differentiation (the so-called spermiogenesis phase) to form highly polarized spermatozoa. Unique characteristics of these cells are featured by the quasi-complete loss of the cytoplasmic content, appearance of the flagella apparatus and drastic size reduction of the nuclear compartment. These major cytological changes give rise to the tiniest mammalian cell type that has the ability to move in order to fulfil its function of delivering to its target, The oocyte, The compacted and, consequently, protected paternal genomic moiety. Up to the spermatid stage the germ cell chromatin presents a somatic organization consisting of short (147 bp) DNA segments wrapped around a histone octamer to form a nucleosome [1]. During spermiogenesis, most (but not all) canonical histone core proteins (H3, H4, H2A, and H2B) are replaced by testis-specific histone variants such as TH2B, H3t, H2AL2 & 5 [2–5]. It is assumed that the inclusion of such variants allows a more dynamic chromatin structure that permits the upcoming changes. Subsequently, histones, both canonical and testicular variants, are largely replaced by small basic proteins called transition nuclear proteins (Tnps), and find themselves replaced by even smaller and more basic proteins called protamines [6,7]. Protamines and DNA organize themselves into a ring-shaped structure called a toroid, containing up to 100 kb of DNA that ultimately piles up along the chromosomes, greatly increasing the level of the DNA compaction [8–11]. This sequence of events allows a strong nuclear and cell size reduction, when compared to any somatic cell [12]. Together with the fact that these modifications enable optimization of cell mobility, they also contribute to passive protection of the paternal sperm genome in anticipation of its long post-testicular journey to the site of fertilization [13].

Another unique feature of this reshaping of the mammalian sperm, chromatin, is that the supra-organization of the chromosomal chromatin is also tightly ordered and conserved from one sperm cell to another. This has led to the observation that chromosomes are not randomly distributed in the sperm nucleus and that they occupy domains, called chromosome territories (CTs) [14–16]. A limited number of species have been investigated, to date, and for those analyzed (mainly human) not all chromosomes were mapped in the sperm nucleus, with the exception of the porcine sperm [14]. The localization of specific chromosomal regions such as telomeres and centromeres were also investigated in the human sperm nucleus [17,18]. As is the case in somatic cells, sperm cell chromosomes are attached to a nuclear protein scaffold, called the sperm nuclear matrix, which consolidates the structure [19–21]. Here too, The manner in which chromosomes are attached to the sperm nuclear matrix is unique to that cell lineage and is dissimilar to the somatic situation [19,22]. Two non-exclusive theories have been proposed to explain the positioning of chromosomes in the nucleus of a somatic cell. The first is "gene density" with the assumption that gene-poor chromosomes orient themselves toward the nuclear periphery while gene-rich chromosomes are located toward the nuclear interior [23,24]. The second theory, and in our opinion the more pertinent, takes chromosome size into account since, at least in the human sperm, it appears that small chromosomes are located in the center of the nucleus while larger chromosomes are located at the periphery [16,25,26]. Whether the human sperm nuclear organization reflects that of other mammals is a matter of debate.

For many years it was reported that mature spermatozoa do contain residual histones and that the quantity of the so-called persisting histones was species-specific. Indeed, it was estimated that about 1–2% of mouse, hamster, and bull sperm DNA was still associated with histones [27–29] and that this value increased to 15% in human sperm [30–34]. First, attributed to an incomplete, therefore deficient, spermiogenesis program, it was recently reported that persisting histones in the sperm nucleus were not random, but were deliberately excluded from the histone-to-protamine exchange. Although, there is a controversy regarding the extent and quality of nucleosome retention in mammalian spermatozoa it is clear that histones are found in large domains punctuating the protamine-toroidal stacks along the chromosomes and, in addition, nucleosomes persist at each small string of DNA, connecting the adjacent toroids [20]. The consensual explanation for this situation is that these particular paternal regions that maintain a somatic-like organization will be more prone to reactivation early after fertilization at the onset of the developmental program. In support of this hypothesis were the observations that the genes important for the early developmental program were found located in such histone-containing regions [30–32], and that the origins of the paternal DNA replication necessary, prior to the first division of segmentation, were located in the short histone-containing DNA segments, connecting the toroids and is attached to the nuclear matrix [19,35–38]. It is thought that this ordered-organization of the paternal chromosomes in the sperm nucleus is essential after fertilization, during the sequential decondensation phase of the male nucleus into the male pronucleus [16,39].

In recent years, we have shown in a mouse model that these histone-rich regions, particularly those that are attached to the nuclear matrix were mainly localized at the sperm nuclear periphery and at the base of the sperm nucleus towards the so-called annulus domain [35,40]. In agreement with the lower level of condensation and the peripheral easy access of these histone-associated DNA domains we also demonstrated that these regions were particularly susceptible to DNA damage and in particular to oxidative DNA damage [35]. We also reported that smaller chromosomes were highly susceptible to DNA oxidation [41] in the mouse sperm nucleus. We demonstrated that this was not related to their content of persisting histones, but rather to the more peripheral and basal position of small chromosomes [36]. These observations led to the conclusion that in contrast to human sperm chromosomal organization, which as mentioned above, showed small chromosomes, located more in the central axis of the sperm nucleus, The situation was different in the mouse. This prompted a more precise analysis of the architecture of the mouse sperm nucleus. In the present study, we used three-dimensional fluorescence in situ hybridization (3D-FISH), confocal microscopy, and computational analysis of 3D structures to analyze the topology of at least twelve mouse sperm chromosomes. This has allowed us to propose the largest map of chromosome territories in murine sperm, to date. Our access to $Gpx5^{-/-}$ transgenic mice, in addition to wild-type controls, allowed us to conduct an analysis of chromatin organization in what now appears to be a frequent type of sperm nuclear damage, i.e., nuclear oxidation [42]. This mouse model was very pertinent to address this question because we reported earlier that $Gpx5^{-/-}$ males present mild oxidative sperm DNA damage that does not translate to an increase in either sperm DNA fragmentation or nuclear decondensation. This transgenic mouse model was particularly interesting, therefore, as it dissociates the effect of severe sperm DNA damage from the low-grade DNA oxidation situation commonly seen in infertile patients. Indeed, we recently demonstrated that males in two-thirds of couples entering an infertility program, showed mild to severe sperm DNA oxidation. Our aims were then to investigate whether chromosomal 3D parameters including volume and surface area would be affected by DNA oxidation.

2. Results

2.1. Localization of Chromosome Territories in Murine Spermatozoa

Previously, we hypothesized that the localization of chromosomes, in the mouse sperm nucleus, could explain their different susceptibility to oxidative damage, as revealed after immunoprecipitation of the oxidized DNA regions, followed by high throughput sequencing approaches [41]. This statement was supported by the fact that we were able to co-localize the smallest murine chromosome (chromosome 19), with a focal point of oxidative DNA damage, in the $Gpx5^{-/-}$ sperm nucleus [41]. To lend support to this statement, we looked at the nuclear distribution of a total of twelve chromosomes (both long and short chromosomes) using the *FISH* assay, in a whole chromosome-painting approach, in both WT and $Gpx5^{-/-}$ sperm nuclei. Figure 1 shows representative confocal microscopy photographs going through the middle of the sperm head for each chromosome investigated. To facilitate this analysis, we arbitrarily divided the mouse sperm head into four distinct areas, as schematized in Figure 1. For each selected chromosome, a minimum of three hundred and fifty sperm cells were analyzed and preferential chromosome positions were determined. It is clear that the small chromosomes, including chromosomes 17, 18, and 19, localized to the basal part of the sperm nucleus, whereas a long chromosome, such as chromosome 1, localized preferentially to the ventral area (see Figure S1, supplemental data). Chromosome 15 and the X and Y sex chromosomes also clearly localized to the dorsal area (Figure 1). Assignation to a preferential domain was easy for these chromosomes because a clear preference was found for these particular locations (see Table 1). In contrast, assignation to a preferential area was more difficult for some chromosomes. For example, two chromosomes (3 and 12) were statistically equally-assigned to two sperm head areas, namely, basal and ventral for chromosome 3 and basal and apical for chromosome 12 (Table 1). Chromosome 2 was peculiar as it was equally localized among the four distinct areas (Table 1). When the same analysis was carried out using $Gpx5^{-/-}$ oxidized sperm, it was clear that no difference was recorded (see Table 1).

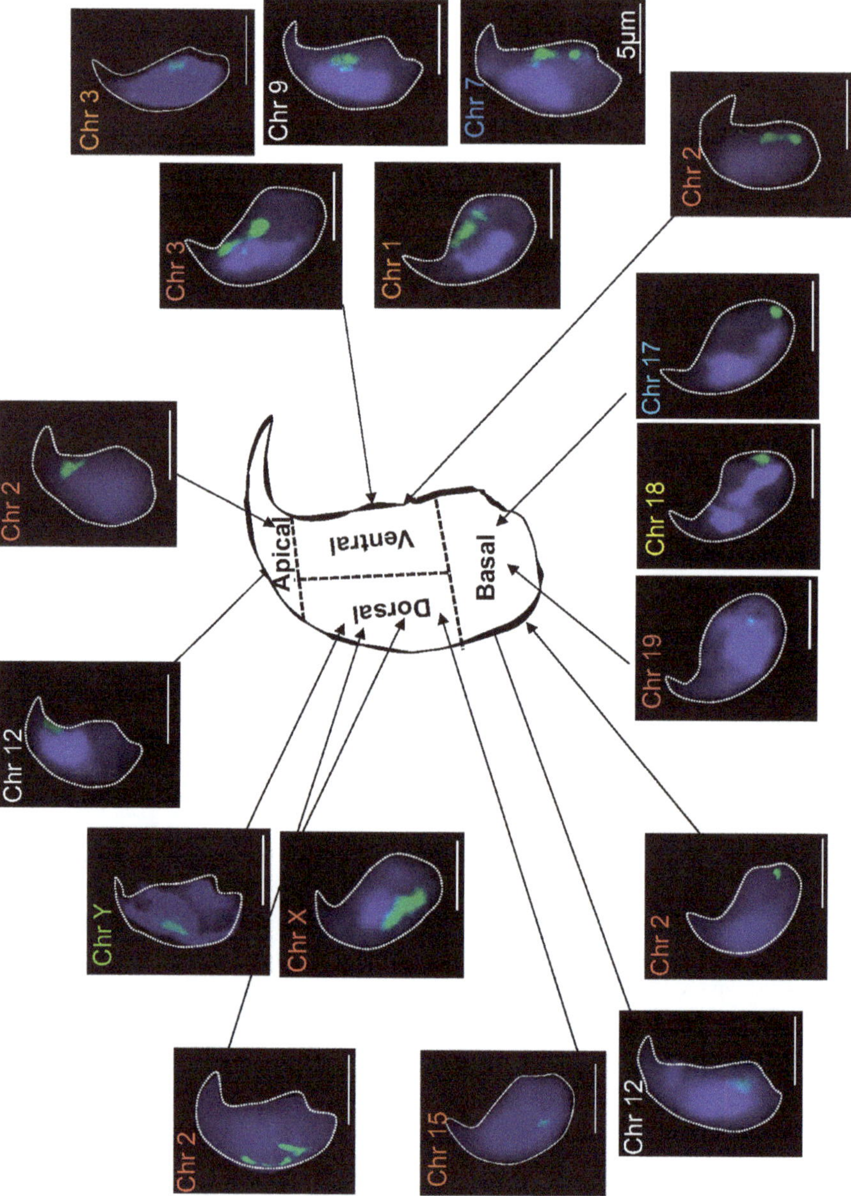

Figure 1. Chromosome mapping in WT mouse sperm nucleus. Schematic representation of a wild-type (WT) mouse sperm nucleus, arbitrarily divided into four regions (apical, dorsal, ventral, and basal). The position of each selected chromosome was detected by fluorescence in situ hybridization (FISH). Green (FITC) staining represents the chromosome position (n = 350 spermatozoa). Nuclei were stained blue with DAPI. Nuclei were captured in Z-stacks by using confocal microscopy and subjected to deconvolution (Huygens software, Hilversum, The Netherlands). Scale bar represents 5 µm (white line). Chr: Chromosome.

Table 1. Regional mapping of chromosomes in WT and $Gpx5^{-/-}$ mouse sperm nuclei. Chromosome positions are assigned, determined in WT and $Gpx5^{-/-}$ mouse sperm nuclei, using FISH. Spermatozoa (n = 350) were counted for each chromosome studied and per genotype. The orange box denote the main position of chromosome.

	WT				$Gpx5-/-$			
	Basal	Apical	Ventral	Dorsal	Basal	Apical	Ventral	Dorsal
Chr 1	27.9	3.4	49.7	19	29.4	8.2	46.3	16.1
Chr 2	25.1	20.2	28.4	26.3	25.5	21	27.5	26
Chr 3	35.9	13.8	31.8	18.5	N.D.			
Chr 7	32.5	9.8	40.3	17.4	29	7.5	47.5	16
Chr 9	29.5	9.8	49	11.7	30	3.8	44.6	21.6
Chr 12	36.8	32.9	13.1	17.2	34.2	27.1	18.9	19.8
Chr 15	21.8	2.8	22.8	52.6	19	7	24	50
Chr 17	57.2	14.1	13.2	15.5	53.8	15.8	16.2	14.2
Chr 18	58.2	22.4	11.2	8.2	57.2	24.3	11.1	7.4
Chr 19	67.2	17	13.6	2.2	61.5	18.4	13.4	6.7
Chr X	7.7	20.7	7.5	64.1	5.3	30.3	4.1	60.3
Chr Y	3.8	29.9	7.2	59.1	4.5	25.4	5.3	64.8

Chr: Chromosome. N.D. not-determined.

Taking advantage of the 3D-reconstructed images we examined two topological parameters (volume and surface area), for each chromosome in the WT genetic background. As shown in supplemental Table S1 and supplemental Figure S1, it is clear that there is a linear relationship between the size of a given chromosome and the volume/surface it occupies in the mouse sperm nucleus. Only chromosome 2 behaved in a peculiar manner, since the linear relationship was validated in only 25% of the analyzed sperm—those in which chromosome 2 localized to the basal area (B in supplemental Table S1 and supplemental Figure S1). Strikingly, when chromosome 2 localized to different areas of the sperm nucleus the linear relationships (volume vs. size and surface vs. size) were lost (supplemental Figure S1). This was particularly true when chromosome 2 was located in the ventral (V) and apical (A) areas and to a lesser extent in the dorsal (D) area. Interestingly, contrasting effects were recorded in these two situations, revealing that when chromosome 2 localized to the ventral and apical areas of the sperm nucleus, its footprint (volume/surface) in the sperm nucleus differed from that when localized to the basal area.

2.2. Centromeres, Telomeres, and Histone-Rich Domains Clustered in the Mouse Sperm Nucleus

Using immunocytochemistry and *FISH*, we further investigated the localization of particular chromosomal subdomains, namely centromeres and telomeres. To do so, we used a pan-centromere specific H3 variant (CENP-A) antibody to detect this ubiquitous centromeric protein (Figure 2A). 3D reconstruction using Imaris software showed that centromeres aligned and clustered along the dorsal and basal ridges of the sperm head (Figure 2B). A similar localization was observed by *FISH* when looking at telomeres (Figure 2C,D) suggesting that in the mouse sperm nucleus, centromeres and telomeres co-localize. No difference in the localization of centromeres and telomeres was recorded when $Gpx5^{-/-}$ sperm nuclei were examined (data not shown). We used three specific histone antibodies (1 canonical and 2 testis-specific variants, respectively, H3, TH2B, and H2A.Z) to corroborate and complete earlier reported partial observations [35] regarding the localization of persisting histones in the mouse sperm nucleus, in immunofluorescence confocal microscopy approaches, associated with 3D Imaris reconstruction. We confirm the basal and dorsal peripheral localization of these persisting histones and their consistently overlapping localization (Figure 3). The 3D Imaris reconstruction, shown in parallel (right panels) in the same Figure, clearly reveals the basal and dorsal ridge localization of these histone-rich domains in what could be designated a "punk-head" distribution. Topoisomerase 2ß, a sperm nuclear matrix protein (Figure 3), as well as the classical cytoskeleton protein ß-tubulin (Figure 3), also fall into these dorsal peripheral and basal ridge domains as was partly shown in the earlier study [30].

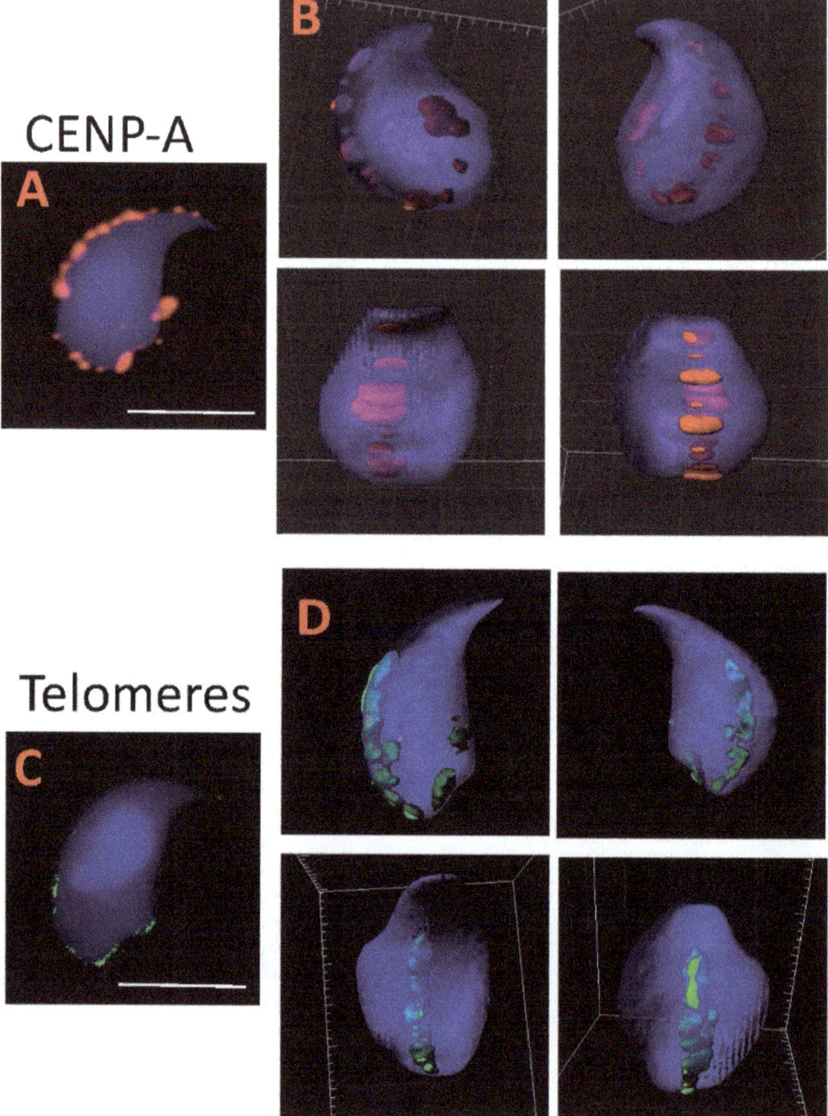

Figure 2. Representative image of telomere and centromere positions in WT mouse sperm nucleus. The centromere-specific histone H3 variant (CENP-A, red (**A**,**B**)) and telomeric probes ((**C**,**D**), red) were used in immunofluorescence or FISH approaches, respectively. Nuclei were stained blue with DAPI. Nuclei were captured in Z-stack, using confocal microscopy, and subjected to deconvolution (Huygens software, Netherlands). The 3D models were obtained with Imaris software (Bitplane, Switzerland). The set of views per staining represented is a representative nucleus from a pool of 30 spermatozoa. Scale bar in confocal images represents 5 μm (white line).

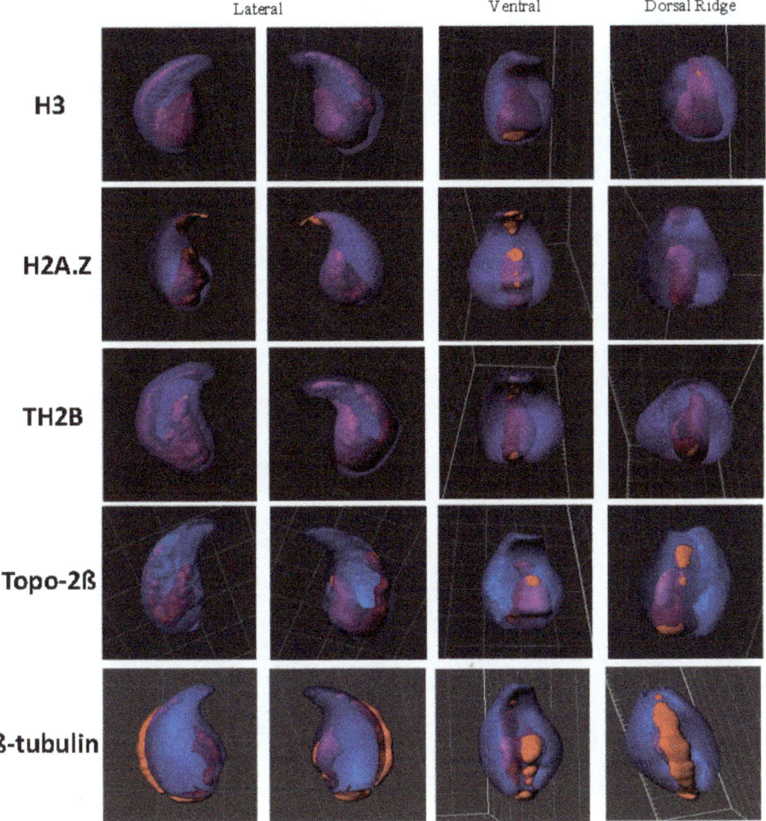

Figure 3. Representative image of chromatin components in WT mouse sperm nucleus. Representative confocal and different views are shown for each component of sperm chromatin in mouse sperm nucleus: Histone H3, histone variant H2A.Z, testis-specific histone variant TH2B, nuclear matrix protein Topoisomerase-II, and ß-tubulin in WT mouse sperm nucleus. Nuclei are captured in Z-stacks using confocal microscopy and subjected to deconvolution (Huygens software, Netherlands). The 3D models were obtained with Imaris software (Bitplane, Switzerland). The set of views per component is a representative nucleus of thirty spermatozoa.

2.3. Oxidative DNA Damage Does Affect 3D-Parameters of the Mouse Sperm Nucleus

Taking advantage of the confocal images and the power of the Imaris software analysis, we looked in more detail at sperm nuclear 3D-parameters, including volume and surface area, comparing WT and $Gpx5^{-/-}$ spermatozoa. An average value for each parameter (volume and surface area) was obtained from each sample and each condition tested (untreated, NaOH- or DTT-treated) by looking at a pool of thirty spermatozoa. The data are presented in Table 2. Untreated WT spermatozoa showed a mean nuclear volume of 66 μm³ and a mean nuclear surface area of 93.9 μm². These parameters were significantly different in $Gpx5^{-/-}$ spermatozoa, which had a mean nuclear volume of 54.8 μm³ ($p < 0.001$) and a mean surface area of 80.2 μm² ($p < 0.001$), revealing a greater state of nuclear condensation. Examination of the detailed shape of the 3D-reconstructed sperm nuclei revealed repeated differences between the WT and $Gpx5^{-/-}$ animals. As shown in Figure 4, with representative photographs of 3D-reconstructed nuclei, $Gpx5^{-/-}$ sperm nuclei present a smoother surface when compared to the more irregular aspect of the WT sperm nuclei. The use of different mild denaturing treatments, namely DTT (2 mM) or NaOH (1.5 N), revealed distinct reactions when WT sperm were compared to $Gpx5^{-/-}$ sperm and confirmed the specific effect of oxidation on the sperm nucleus. As presented in Table 2, when NaOH was used to produce a mild denaturation of the sperm chromatin (by classical breakage effects on the hydrogen bonds linking DNA base pairs), we recorded/*observed* a significant increase in sperm nuclear volume and surface area, in both genetic backgrounds (WT and $Gpx5^{-/-}$). However, $Gpx5^{-/-}$ sperm nuclei remained more condensed than WT following treatment with alkali. In contrast, when DTT (a non-ionic detergent that specifically reduces disulfide bonds to free thiols) was used, we observed a marked effect on both sperm nuclear volume and surface area, in the $Gpx5^{-/-}$ mice, as compared with WT controls (Table 2). This is in agreement with the idea that although $Gpx5^{-/-}$ sperm nuclei appear more condensed, they also appear to be significantly less robust when exposed to a mild, reducing environment. These differences in the nuclear reactivity of oxidized or non-oxidized sperm nuclei, when exposed to mild denaturing conditions, can be visualized, as shown in Figure 4. In panel C (Figure 4C), when no denaturing treatment was performed, The $Gpx5^{-/-}$ sperm nuclei presented the smooth aspect, as noted above. When NaOH was used as a mild denaturing treatment, there was no significant change regarding the smooth shape of the sperm nuclei in either genetic background (Figure 4B). However, when mild denaturation was carried out with DTT, it was obvious that the $Gpx5^{-/-}$ sperm nuclei then presented a dense granular aspect (Figure 4C) that was not observed in the WT.

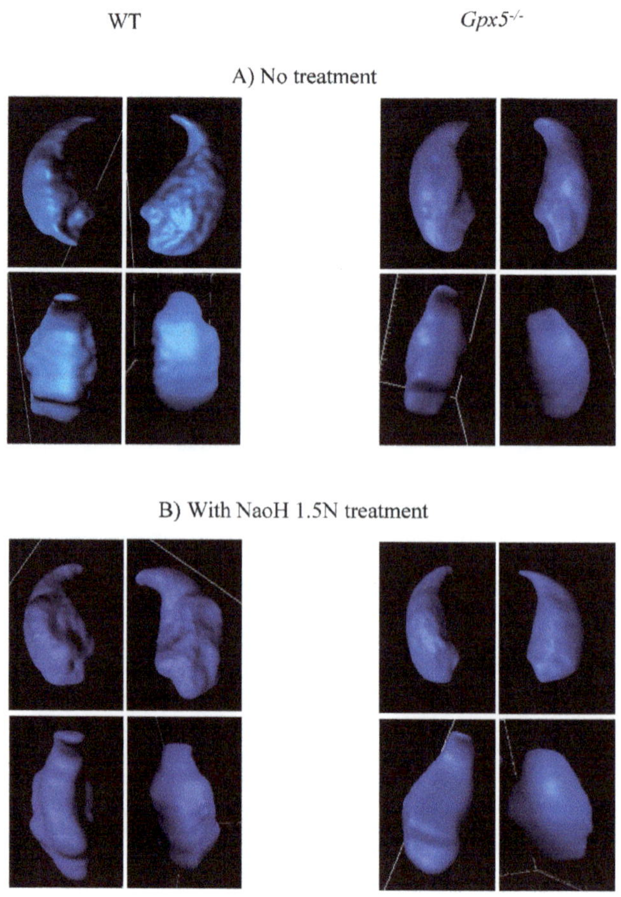

Figure 4. *Cont.*

C) With DTT treatment (2mM 45min)

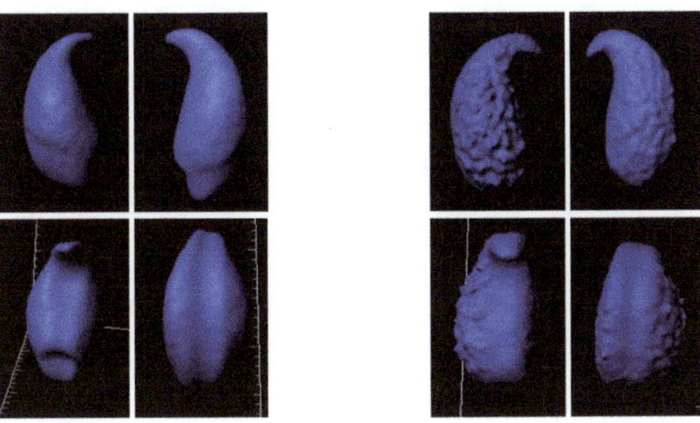

Figure 4. NaOH-mediated or DTT-mediated mild denaturation provokes distinct effects on the WT and the *Gpx5*$^{-/-}$ nuclei.

Table 2. Three-dimensional parameters of sperm nuclei according to treatment and genotype.

	WT	***Gpx5*$^{-/-}$**
Average volume (μm^3)	66	54.8 [a]
Average Area (μm^2)	93.9	80.2 [a]
Nucleus with NaOH 1.5N Treatment	**WT**	***Gpx5*$^{-/-}$**
Average volume (μm^3)	109 [d]	93.5 [b,d]
Average Area (μm^2)	138 [d]	113.5 [b,d]
Nucleus with DTT Treatment (2 mM, 45 min)	**WT**	***Gpx5*$^{-/-}$**
Average volume (μm^3)	85.3 [d]	130.4 [c,d]
Average Area (μm^2)	108.4 [d]	143.3 [c,d]

Volume and surface area of nuclei were calculated from 3D photographs obtained of Z-stack images, generated with the Imaris software (Bitplane, Switzerland). Nuclei were captured in Z-stacks, using confocal microscopy and subjected to deconvolution (Huygens software, Hilversum, The Netherlands). The resulting distribution of the different parameters are shown in the table for each genotype (WT and *Gpx5*$^{-/-}$). The mean was calculated on thirty spermatozoa per condition. DTT: Dithiothreitol; NaOH: Sodium hydroxide. 'a' represents $p < 0.001$ for WT no treatment condition; 'b' represents $p < 0.001$ for WT NaOH condition; 'c' represents $p < 0.001$ for WT DTT condition; 'd' represents $p < 0.001$ for no treatment/genotype condition.

3. Discussion

In recent years, it has become apparent that mammalian sperm nucleus organization has implications for fertilization and early embryogenesis [14,15,43–46]. It was shown, mainly in human spermatozoa, that most chromosomes occupy discrete and well-defined territories in a polar/radial distribution that could be partly related to their size [44–46], The shape/volume of the mature sperm cell and the kinetics of the oocyte-driven decondensation program of the paternal nucleus post-fertilization [47,48]. How this highly-ordered organization of the sperm chromatin is achieved, controlled, and maintained in each sperm cell, throughout spermiogenesis and beyond, is still largely unknown. Whether the human sperm chromatin organization applies to murine sperm and how susceptible this organization is to mild nuclear and DNA damage, as represented by the common situation of sperm DNA oxidative damage, are questions we addressed in this study.

Using FISH experiments, we determined the position of a total of twelve chromosomes in the mouse sperm nucleus. Both short and long autosomes and the two sex chromosomes were analyzed. As reported for the human sperm nucleus, and suggested for other species (including mouse, bovine, pig, and rat), using a smaller subset of chromosome probes when compared to the present work [14–17,45,49–56], chromosome positions in the mouse sperm nucleus were not random. This situation seems to be confined to mammals since a tandem head-to-tail organization of sperm chromosomes, in a defined order, was observed in monotremes and marsupials [57,58] while no particular organization was detected in non-mammals, including chicken and planarian spermatozoa [59,60].

Due to this peculiar, asymmetric hook-shape morphology of the mouse sperm head it was difficult to use a polar/radial axis to map the mouse sperm head, as has been performed in other species [15]. We arbitrarily separated the mouse sperm head into four compartments (apical/basal/dorsal/ventral), while still permitting comparative analyses with other species. In the mouse, smaller chromosomes were found to occupy a basal localization, whereas longer chromosomes were preferentially found in the ventral area with the sex chromosomes located in the dorsal area of the sperm nucleus. This appears to be distinct from the human situation since it was shown that small autosomes as well as sex chromosomes occupy a rather central position in human sperm [16]. Some of the CTs appear to be small while others are larger. Our assumption is that it is both related to the respective size of the chromosomes (since we did observe that there is a positive correlation between the size and the volume of the chromosome, as shown in supplementary Table S1). However, it could also be partly related to the number of times by which the chromosomes—which are folded to fit into the tiny nuclear volume—are longer. Although a preferential position could be assigned for most of the chromosomes examined, this did not hold for all chromosomes. Four chromosomes (chr 3, 7, 9, and 12) were equally assigned to two distinct areas, while one chromosome (chr 2) was very plastic and was found evenly distributed among the four arbitrarily-defined nuclear areas. For those chromosomes that were equally distributed between the two distinct nuclear domains, one explanation could arise from the fact that statistically one out of two spermatozoa examined was either a Y-spermatozoon or an X-spermatozoon. The size difference between the sex chromosomes (both localized in the dorsal area) could explain the alternate positions of these autosomes. This hypothesis is strengthened by the observation that overall Y-sperm and X-sperm show a similar nuclear volume (not shown here) suggesting that the necessary adjustment to accommodate the X or Y chromosome size-difference does not rely on nuclear volume variation. Furthermore, when looking at individual chromosome 3D-parameters (i.e., volume and surface area) we observed that chromosomes 3 and 12 (two chromosomes that show equal occupancy of two distinct locations, basal or ventral for chromosome 3 and basal or apical for chromosome 12) have the same footprint, irrespective of their location. This suggests that the nuclear space adjustment necessary to accommodate the X or the Y chromosome does not rely on different folding of individual chromosomes, but rather on different chromosome positions. These hypotheses would require verification using a triad-detection system with probes targeting a chosen autosome, together with probes targeting sex chromosomes. Chromosome 2 is rather intriguing as it distributes equally in any of the four arbitrarily defined nuclear areas. This observation is not unique to murine sperm, since human sperm chromosome 13 showed identical behavior [17]. Although a rather long autosome, it seems that chromosome 2 is considered as an adjustment variable in the mouse sperm nucleus. In addition, we and others have data suggesting that mouse chromosome 2 is a rather accessible chromosome in the mouse sperm nucleus, since it was observed on several occasions that, when purifying murine sperm DNA for high throughput sequencing strategies, one systematically obtained a large excess of chromosome 2 sequences in comparison to other chromosomes [41,61,62]. This suggests a peripheral localization of this chromosome in the mouse sperm nucleus as it does not appear to be less-condensed than other autosomes [41].

Telomeres have recently been assigned a chromosome stabilizing function that is important for reproduction [63] and it is proposed that telomeres are the first chromosomal regions to respond to oocyte decondensing factors that lead to the formation of the male pronucleus [46]. As suggested earlier in mouse sperm nucleus [49,56] and recently confirmed for the human sperm nucleus [17,52,56,64], we showed here that telomeres in murine sperm are also organized in clusters located at the periphery of the sperm nucleus in an edge-like/ridge-like manner, starting from the base of the nucleus and extending along the dorsal side, in close proximity to the peripheral nuclear matrix. With regard to centromeres, another characteristic domain of chromosomes rich in repeated sequences we found that in the mouse sperm nucleus, they were also located in clusters, at the periphery, with the same edge-like/ridge-like organization. In a previous study, it was shown via *FISH* that the distribution of centromeres in testicular sperm (not fully mature) are clustered at the surface of the heterochromatic chromocenter (schematic representation in Figure 5B). This differed from our study [49] in which fully mature post-testicular (i.e., epididymal) sperm were evaluated. An organization similar to the one we report here was recently described in human sperm nuclei, in which the centromeres were distributed as single clusters [64]. The present localization of centromeres in murine sperm, determined by using the histone H3 variant CENP-A, is in agreement with previous data reporting that histones in mature murine sperm are preferentially located in the basal and dorsal peripheral areas of the nucleus [35]. In view of these results we propose a new model for telomere and centromere organization in murine sperm nuclei (Figure 5C). It would appear that in the mouse, both telomeres and centromeres are closely located at these dorsal peripheral and basal nuclear domains that were shown elsewhere (as well as here) to be domains rich in nuclear matrix attachment components [30] and rich in histone [65,66]. It is interesting to note that the paternal DNA associated with these nuclear regions was shown to be important both for male pronucleus formation and for the first round of DNA replication [19,37] which are early events of embryo development. As it has been well described in a recent review [67], The organization of the sperm nucleus seems to be an important factor for male fertility and embryo development that will require further analysis.

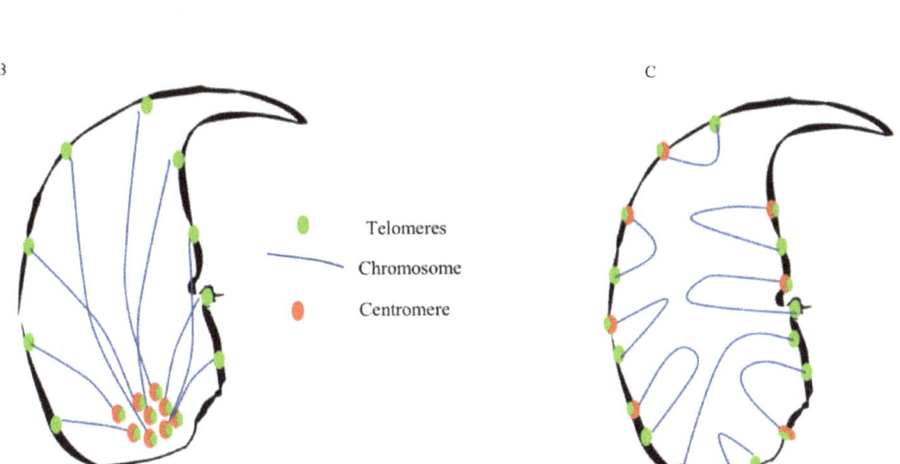

Figure 5. Schematic representation of the proposed models of telomere and centromere organization within murine sperm nuclei. Panel (**A**) presents a schematic representation of the murine acrocentric chromosome with two telomere regions (green) at either end of the chromosome and one centromere (red). Panel (**B**) presents a schematic representation of the murine chromosome model in which the centromeres (red) gather in a chromocenter, with the chromosome (light blue) stretching out toward the telomere (green) localized at the peripheral region. In panel (**C**), we present a refined version of the model, based on our observations, which depicts a more segmented organization, with localization of telomeres (green) and centromere (red), throughout the murine nucleus.

Concerning the susceptibility of the sperm chromatin organization to oxidative alterations, gross examination of the nuclear topology of the chromosomes (studied in this work) shows that they are unaffected by the mild oxidative environment present in the $Gpx5^{-/-}$ transgenic mouse strain. This is supported by the fact that we did not record significant differences in the distribution of the chromosomes, in the four arbitrarily defined regions, when comparing WT and transgenic sperm (Table 1). However, a recent study did show that high levels of DNA damage in human sperm (such as significant DNA fragmentation) could disrupt the position of the centromeres [68]. This suggested that chromosome 3D organization may be impacted depending on the level of sperm DNA damage.

When looking at nuclear 3D-parameters, such as nuclear volume and surface area we confirmed, as expected, The susceptibility of the nucleus to oxidative alterations. This is evidenced by the observations that both nuclear volume and nuclear surface area are significantly diminished in the $Gpx5^{-/-}$ spermatozoa, when compared with WT sperm. This is in line with the idea that when the epididymis-secreted GPx5 protein is absent, it leaves more luminal H_2O_2 that is used by the sperm-nucleus GPx4 (acting here as a disulphide isomerase) to generate disulphide bridges between the sperm nuclear protamines, leading to a greater state of nuclear condensation [69]. The observation, after the 3D-reconstruction, that the $Gpx5^{-/-}$ sperm show a smoother nuclear surface when compared to the WT sperm which has a "goose-bumps" aspect, is interesting as it distinguishes nuclear domains responding differentially to this oxidation-mediated increased condensation. In particular, The use of different, mild, denaturing treatments (alkaline *versus* reductive denaturation) emphasized the point that even though a mildly oxidized sperm nucleus may appear well-condensed (as for $Gpx5^{-/-}$ spermatozoa) it is highly susceptible to mild reductive conditions. This is important in clinical practice as clinicians may be misled when using assays such as the aniline blue or the toluidine blue, to determine the level of sperm nuclear condensation as an indicator of sperm nuclear integrity. Therefore, The type of mild denaturation technique chosen will be critical to correctly determine the level of sperm nuclear integrity. These considerations support our credo that a solid evaluation of sperm nuclear integrity/solidity, prior to assisted reproductive technology (ART), should include several additional tests, addressing the issue of DNA fragmentation, DNA oxidation, and nuclear solidity.

4. Materials and Methods

4.1. Ethics Statement

Ethics statement: The present study was approved by the Regional Ethics Committee of Animal Experimentation (CEMEA-Auvergne; Authorization CE99-12) and adhered to the current legislation on animal experimentation in France.

4.2. Animals

The $Gpx5^{-/-}$ mice were derived, as described originally, from the C57BL/6 genetic line [41,42]. Mice used in this study (eight mice per genotype) were maintained and housed in temperature-controlled rooms with 12-h light/dark cycles. Mice had ad libitum access to food and water. Nine-month-old mice were culled by cervical dislocation and spermatozoa were collected from the caudal segment of the epididymis.

4.3. Immunocytochemistry and Fluorescence in situ Hybridization (FISH) Assays

All immunofluorescence procedures were performed as previously described [35]. Briefly, spermatozoa were resuspended in a decondensing buffer (2 mM DTT and 0.5% triton X-100 in PBS) and incubated for 45 min, at room temperature. After centrifugation at $500\times g$ for 5 min, at room temperature, spermatozoa were resuspended in PBS, numbered, and deposited onto a glass plate. For FISH assays, spermatozoa were recovered as described previously [41]. A fraction aliquot of 10×10^6 spz/mL was centrifuged at $560\times g$, for 5 min and re-suspended in 1.25 mL fresh Carnoy's fixative (3:1 ethanol:acetic acid). This spermatozoa-containing solution was spread on the slides (up to 25,000 spermatozoa/slide) then slides were dried for 1 h, at room temperature (RT), and stored at $-20\,°C$ (Superfrost® slides, Thermo Fisher Scientific, Illkirch, France). After 24 h, slides were defrosted at RT and placed in a coplin jar with saline-sodium citrate solution 2X (SSC 2X), for 15 min at $37\,°C$. Slides were dried for 5 min, at RT, and denatured using NaOH 1.5 N (1 min). Slides were then incubated in a coplin jar with SSC 2X for 30 min, at $70\,°C$ ($\pm1\,°C$). The coplin jar was left at RT. Slides were successively incubated for 1 min in SSC 0.1X at RT, NaOH 0.07 N at RT, SSC 0.1X at $4\,°C$, and SSC 2X at $4\,°C$. Slides were transferred through a series of ethanol washes for 1 min, each starting with 70%, 95%, and finally 100% ethanol. Slides were left to dry at RT. DNA probes were applied to a sterile coverslip, pre-warmed at $37\,°C$, and sealed using paraffin. Finally, slides were incubated in a dark humidified chamber at $37\,°C$, for 48 h. Mouse chromosome-painting probes (Metasystems, Altlussheim, Germany) and telomere probes (Panagene, Altlussheim, Korea) were prepared according to the manufacturer's instructions. After a 48-h incubation period, The slides were washed with SSC 0.4X for 2 min, at $70\,°C$ ($\pm1\,°C$), and 30 s, in SSC 2X, with Tween 0.05%, and for two successive rinses. Vectashield® with DAPI (Vector Laboratories) was added to each slide to counterstain the sperm cell nucleus. Finally, coverslips were mounted, sealed, and slides were stored in the dark at $-20\,°C$, until observation.

4.4. Microscopy

Confocal Z-stacks were captured using a Leica SPE confocal microscope (Leica Microsystems, Wetzlar, Germany) and a $40\times$ oil immersion objective was used for all acquisitions. At least eighty stacks per nucleus were captured and the distance between Z stacks was 0.21 µm. Chromosome territory was assigned after counting not less than three hundred and fifty spermatozoa per chromosome and the percent of spermatozoa presenting a chromosome at one or more given positions was established. A Zeiss microscope Axioplan2 (Carl Zeiss, Oberkochen, Germany) was used to perform these observations.

4.5. Image Analysis Measurements of 3D Parameters

All the images were deconvoluted using Huygens software (Scientific Volume Imaging, The Netherlands) before analysis. Spermatozoa volume and surface area were measured using Imaris Version 7.6 software (Bitplane AG, Zurich, Switzerland). The mean of each parameter was calculated with at least 30 spermatozoa.

4.6. Statistics

Mann-Whitney and Spearman correlation analyses were performed using GraphPad Prism® software. The difference was considered significant when $p < 0.001$ (**).

Supplementary Materials: The following are available online at http://www.mdpi.com/2073-4425/9/10/501/s1, Figure S1: Correlation of volume/surface area and size, Table S1: Three-dimensional parameters of sperm chromosomes in WT mouse 3 sperm nucleus.

Author Contributions: Authors A.C. and A.K. were involved in the conception and design of the experimental manipulations, assisted with the molecular techniques, and wrote the manuscript. C.D.-S., C.G. and S.B. contributed to the experimental manipulations. J.D. drafted and critically reviewed the manuscript. J.H.-B., R.G. and F.S. also performed a critical review of the manuscript. All authors have read and approved the final manuscript.

Funding: This research was funded by the CNRS, INSERM & French Ministry of Higher Education and Research (MESR) through operating grants attributed to JRD.

Acknowledgments: The authors would like to thank the CNRS, INSERM, and UCA for their financial support, as well as the former *Région Auvergne* for contribution to this research.

Conflicts of Interest: The authors declare no conflicts of interest.

References

1. Champroux, A.; Torres-Carreira, J.; Gharagozloo, P.; Drevet, J.R.; Kocer, A. Mammalian sperm nuclear organization: Resiliencies and vulnerabilities. *Basic Clin. Androl.* **2016**, *26*, 17. [CrossRef] [PubMed]
2. Montellier, E.; Boussouar, F.; Rousseaux, S.; Zhang, K.; Buchou, T.; Fenaille, F.; Shiota, H.; Debernardi, A.; Héry, P.; Curtet, S.; et al. Chromatin-to-nucleoprotamine transition is controlled by the histone H2B variant TH2B. *Genes Dev.* **2013**, *27*, 1680–1692. [CrossRef] [PubMed]
3. González-Romero, R.; Méndez, J.; Ausió, J.; Eirín-López, J.M. Quickly evolving histones, nucleosome stability and chromatin folding: All about histone H2A.Bbd. *Gene* **2008**, *413*, 1–7. [CrossRef] [PubMed]
4. Govin, J.; Escoffier, E.; Rousseaux, S.; Kuhn, L.; Ferro, M.; Thévenon, J.; Catena, R.; Davidson, I.; Garin, J.; Khochbin, S.; et al. Pericentric heterochromatin reprogramming by new histone variants during mouse spermiogenesis. *J. Cell Biol.* **2007**, *176*, 283–294. [CrossRef] [PubMed]
5. Hoghoughi, N.; Barral, S.; Vargas, A.; Rousseaux, S.; Khochbin, S. Histone variants: Essential actors in the male genome programing. *J. Biochem.* **2017**. [CrossRef] [PubMed]
6. Balhorn, R. The protamine family of sperm nuclear proteins. *Genome Biol.* **2007**, *8*, 227. [CrossRef] [PubMed]
7. Rathke, C.; Baarends, W.M.; Awe, S.; Renkawitz-Pohl, R. Chromatin dynamics during spermiogenesis. *Biochim. Biophys. Acta* **2014**, *1839*, 155–168. [CrossRef] [PubMed]
8. Allen, M.J.; Bradbury, E.M.; Balhorn, R. AFM analysis of DNA-protamine complexes bound to mica. *Nucleic Acids Res.* **1997**, *25*, 2221–2226. [CrossRef] [PubMed]
9. Brewer, L.; Corzett, M.; Lau, E.Y.; Balhorn, R. Dynamics of protamine 1 binding to single DNA molecules. *J. Biol. Chem.* **2003**, *278*, 42403–42408. [CrossRef] [PubMed]
10. Hud, N.V.; Allen, M.J.; Downing, K.H.; Lee, J.; Balhorn, R. Identification of the elemental packing unit of DNA in mammalian sperm cells by atomic force microscopy. *Biochem. Biophys. Res. Commun.* **1993**, *193*, 1347–1354. [CrossRef] [PubMed]
11. Hud, N.V.; Downing, K.H.; Balhorn, R. A constant radius of curvature model for the organization of DNA in toroidal condensates. *Proc. Natl. Acad. Sci. USA* **1995**, *92*, 3581–3585. [CrossRef] [PubMed]
12. Ward, W.S.; Coffey, D.S. DNA packaging and organization in mammalian spermatozoa: Comparison with somatic cells. *Biol. Reprod.* **1991**, *44*, 569–574. [CrossRef] [PubMed]
13. Champroux, A.; Cocquet, J.; Henry-Berger, J.; Drevet, J.R.; Kocer, A. A Decade of Exploring the Mammalian Sperm Epigenome: Paternal Epigenetic and Transgenerational Inheritance. *Front. Cell Dev. Biol.* **2018**, *6*, 50. [CrossRef] [PubMed]
14. Foster, H.A.; Abeydeera, L.R.; Griffin, D.K.; Bridger, J.M. Non-random chromosome positioning in mammalian sperm nuclei, with migration of the sex chromosomes during late spermatogenesis. *J. Cell Sci.* **2005**, *118*, 1811–1820. [CrossRef] [PubMed]
15. Millan, N.M.; Lau, P.; Hann, M.; Ioannou, D.; Hoffman, D.; Barrionuevo, M.; Maxson, W.; Ory, S.; Tempest, H.G. Hierarchical radial and polar organisation of chromosomes in human sperm. *Chromosom. Res. Int. J. Mol. Supramol. Evol. Asp. Chromosom. Biol.* **2012**, *20*, 875–887. [CrossRef] [PubMed]
16. Zalensky, A.; Zalenskaya, I. Organization of chromosomes in spermatozoa: An additional layer of epigenetic information? *Biochem. Soc. Trans.* **2007**, *35*, 609–611. [CrossRef] [PubMed]
17. Hazzouri, M.; Rousseaux, S.; Mongelard, F.; Usson, Y.; Pelletier, R.; Faure, A.K.; Vourc'h, C.; Sèle, B. Genome organization in the human sperm nucleus studied by FISH and confocal microscopy. *Mol. Reprod. Dev.* **2000**, *55*, 307–315. [CrossRef]
18. Zalensky, A.O.; Breneman, J.W.; Zalenskaya, I.A.; Brinkley, B.R.; Bradbury, E.M. Organization of centromeres in the decondensed nuclei of mature human sperm. *Chromosoma* **1993**, *102*, 509–518. [CrossRef] [PubMed]
19. Shaman, J.A.; Yamauchi, Y.; Ward, W.S. Function of the sperm nuclear matrix. *Arch. Androl.* **2007**, *53*, 135–140. [CrossRef] [PubMed]
20. Ward, W.S. Function of sperm chromatin structural elements in fertilization and development. *Mol. Hum. Reprod.* **2010**, *16*, 30–36. [CrossRef] [PubMed]
21. Gawecka, J.E.; Ribas-Maynou, J.; Benet, J.; Ward, W.S. A model for the control of DNA integrity by the sperm nuclear matrix. *Asian J. Androl.* **2015**, *17*, 610–615. [PubMed]
22. Yamauchi, Y.; Shaman, J.A.; Ward, W.S. Non-genetic contributions of the sperm nucleus to embryonic development. *Asian J. Androl.* **2011**, *13*, 31–35. [CrossRef] [PubMed]

23. Boyle, S.; Gilchrist, S.; Bridger, J.M.; Mahy, N.L.; Ellis, J.A.; Bickmore, W.A. The spatial organization of human chromosomes within the nuclei of normal and emerin-mutant cells. *Hum. Mol. Genet.* **2001**, *10*, 211–219. [CrossRef] [PubMed]

24. Croft, J.A.; Bridger, J.M.; Boyle, S.; Perry, P.; Teague, P.; Bickmore, W.A. Differences in the localization and morphology of chromosomes in the human nucleus. *J. Cell Biol.* **1999**, *145*, 1119–1131. [CrossRef] [PubMed]

25. Bolzer, A.; Kreth, G.; Solovei, I.; Koehler, D.; Saracoglu, K.; Fauth, C.; Müller, S.; Eils, R.; Cremer, C.; Speicher, M.R.; et al. Three-dimensional maps of all chromosomes in human male fibroblast nuclei and prometaphase rosettes. *PLoS Biol.* **2005**, *3*, e157. [CrossRef] [PubMed]

26. Sun, H.B.; Shen, J.; Yokota, H. Size-dependent positioning of human chromosomes in interphase nuclei. *Biophys. J.* **2000**, *79*, 184–190. [CrossRef]

27. Bench, G.S.; Friz, A.M.; Corzett, M.H.; Morse, D.H.; Balhorn, R. DNA and total protamine masses in individual sperm from fertile mammalian subjects. *Cytometry* **1996**, *23*, 263–271. [CrossRef]

28. Balhorn, R.; Gledhill, B.L.; Wyrobek, A.J. Mouse sperm chromatin proteins: Quantitative isolation and partial characterization. *Biochemistry* **1977**, *16*, 4074–4080. [CrossRef] [PubMed]

29. Tovich, P.R.; Oko, R.J. Somatic histones are components of the perinuclear theca in bovine spermatozoa. *J. Biol. Chem.* **2003**, *278*, 32431–32438. [CrossRef] [PubMed]

30. Arpanahi, A.; Brinkworth, M.; Iles, D.; Krawetz, S.A.; Paradowska, A.; Platts, A.E.; Saida, M.; Steger, K.; Tedder, P.; Miller, D. Endonuclease-sensitive regions of human spermatozoal chromatin are highly enriched in promoter and CTCF binding sequences. *Genome Res.* **2009**, *19*, 1338–1349. [CrossRef] [PubMed]

31. Erkek, S.; Hisano, M.; Liang, C.; Gill, M.; Murr, R.; Dieker, J.; Schübeler, D.; van der Vlag, J.; Stadler, M.B.; Peters, A.H.F.M. Molecular determinants of nucleosome retention at CpG-rich sequences in mouse spermatozoa. *Nat. Struct. Mol. Biol.* **2013**, *20*, 868–875. [CrossRef] [PubMed]

32. Hammoud, S.S.; Nix, D.A.; Zhang, H.; Purwar, J.; Carrell, D.T.; Cairns, B.R. Distinctive chromatin in human sperm packages genes for embryo development. *Nature* **2009**, *460*, 473–478. [CrossRef] [PubMed]

33. Gatewood, J.M.; Cook, G.R.; Balhorn, R.; Schmid, C.W.; Bradbury, E.M. Isolation of four core histones from human sperm chromatin representing a minor subset of somatic histones. *J. Biol. Chem.* **1990**, *265*, 20662–20666. [PubMed]

34. Brykczynska, U.; Hisano, M.; Erkek, S.; Ramos, L.; Oakeley, E.J.; Roloff, T.C.; Beisel, C.; Schübeler, D.; Stadler, M.B.; Peters, A.H. Repressive and active histone methylation mark distinct promoters in human and mouse spermatozoa. *Nat. Struct. Mol. Biol.* **2010**, *17*, 679–687. [CrossRef] [PubMed]

35. Noblanc, A.; Damon-Soubeyrand, C.; Karrich, B.; Henry-Berger, J.; Cadet, R.; Saez, F.; Guiton, R.; Janny, L.; Pons-Rejraji, H.; Alvarez, J.G.; et al. DNA oxidative damage in mammalian spermatozoa: Where and why is the male nucleus affected? *Free Radic. Biol. Med.* **2013**, *65*, 719–723. [CrossRef] [PubMed]

36. Zalensky, A.O.; Siino, J.S.; Gineitis, A.A.; Zalenskaya, I.A.; Tomilin, N.V.; Yau, P.; Bradbury, E.M. Human testis/sperm-specific histone H2B (hTSH2B). Molecular cloning and characterization. *J. Biol. Chem.* **2002**, *277*, 43474–43480. [CrossRef] [PubMed]

37. Shaman, J.A.; Yamauchi, Y.; Ward, W.S. The sperm nuclear matrix is required for paternal DNA replication. *J. Cell. Biochem.* **2007**, *102*, 680–688. [CrossRef] [PubMed]

38. Anachkova, B.; Djeliova, V.; Russev, G. Nuclear matrix support of DNA replication. *J. Cell. Biochem.* **2005**, *96*, 951–961. [CrossRef] [PubMed]

39. Solov'eva, L.; Svetlova, M.; Bodinski, D.; Zalensky, A.O. Nature of telomere dimers and chromosome looping in human spermatozoa. *Chromosom. Res. Int. J. Mol. Supramol. Evol. Asp. Chromosom. Biol.* **2004**, *12*, 817–823. [CrossRef] [PubMed]

40. Ward, W.S.; Coffey, D.S. Identification of a sperm nuclear annulus: A sperm DNA anchor. *Biol. Reprod.* **1989**, *41*, 361–370. [CrossRef] [PubMed]

41. Kocer, A.; Henry-Berger, J.; Noblanc, A.; Champroux, A.; Pogorelcnik, R.; Guiton, R.; Janny, L.; Pons-Rejraji, H.; Saez, F.; Johnson, G.D.; et al. Oxidative DNA damage in mouse sperm chromosomes: Size matters. *Free Radic. Biol. Med.* **2015**, *89*, 993–1002. [CrossRef] [PubMed]

42. Chabory, E.; Damon, C.; Lenoir, A.; Kauselmann, G.; Kern, H.; Zevnik, B.; Garrel, C.; Saez, F.; Cadet, R.; Henry-Berger, J.; et al. Epididymis seleno-independent glutathione peroxidase 5 maintains sperm DNA integrity in mice. *J. Clin. Investig.* **2009**, *119*, 2074–2085. [CrossRef] [PubMed]

43. Ioannou, D.; Miller, D.; Griffin, D.K.; Tempest, H.G. Impact of sperm DNA chromatin in the clinic. *J. Assist. Reprod. Genet.* **2016**, *33*, 157–166. [CrossRef] [PubMed]

44. Mudrak, O.S.; Nazarov, I.B.; Jones, E.L.; Zalensky, A.O. Positioning of chromosomes in human spermatozoa is determined by ordered centromere arrangement. *PLoS ONE* **2012**, *7*, e52944. [CrossRef] [PubMed]

45. Zalenskaya, I.A.; Zalensky, A.O. Non-random positioning of chromosomes in human sperm nuclei. *Chromosom. Res. Int. J. Mol. Supramol. Evol. Asp. Chromosom. Biol.* **2004**, *12*, 163–173. [CrossRef]

46. Zalenskaya, I.A.; Bradbury, E.M.; Zalensky, A.O. Chromatin structure of telomere domain in human sperm. *Biochem. Biophys. Res. Commun.* **2000**, *279*, 213–218. [CrossRef] [PubMed]

47. Liu, D.Y.; Baker, H.W.G. Human sperm bound to the zona pellucida have normal nuclear chromatin as assessed by acridine orange fluorescence. *Hum. Reprod.* **2007**, *22*, 1597–1602. [CrossRef] [PubMed]

48. Ioannou, D.; Griffin, D.K. Male fertility, chromosome abnormalities, and nuclear organization. *Cytogenet. Genome Res.* **2011**, *133*, 269–279. [CrossRef] [PubMed]

49. Haaf, T.; Ward, D.C. Higher order nuclear structure in mammalian sperm revealed by in situ hybridization and extended chromatin fibers. *Exp. Cell Res.* **1995**, *219*, 604–611. [CrossRef] [PubMed]

50. Jennings, C.; Powell, D. Genome organisation in the murine sperm nucleus. *Zygote Camb. Engl.* **1995**, *3*, 123–131. [CrossRef]

51. Luetjens, C.M.; Payne, C.; Schatten, G. Non-random chromosome positioning in human sperm and sex chromosome anomalies following intracytoplasmic sperm injection. *Lancet Lond. Engl.* **1999**, *353*, 1240. [CrossRef]

52. Meyer-Ficca, M.; Müller-Navia, J.; Scherthan, H. Clustering of pericentromeres initiates in step 9 of spermiogenesis of the rat (Rattus norvegicus) and contributes to a well defined genome architecture in the sperm nucleus. *J. Cell Sci.* **1998**, *111 Pt 10*, 1363–1370.

53. Powell, D.; Cran, D.G.; Jennings, C.; Jones, R. Spatial organization of repetitive DNA sequences in the bovine sperm nucleus. *J. Cell Sci.* **1990**, *97 Pt 1*, 185–191.

54. Sbracia, M.; Baldi, M.; Cao, D.; Sandrelli, A.; Chiandetti, A.; Poverini, R.; Aragona, C. Preferential location of sex chromosomes, their aneuploidy in human sperm, and their role in determining sex chromosome aneuploidy in embryos after ICSI. *Hum. Reprod. Oxf. Engl.* **2002**, *17*, 320–324. [CrossRef]

55. Tilgen, N.; Guttenbach, M.; Schmid, M. Heterochromatin is not an adequate explanation for close proximity of interphase chromosomes 1–Y, 9–Y, and 16–Y in human spermatozoa. *Exp. Cell Res.* **2001**, *265*, 283–287. [CrossRef] [PubMed]

56. Zalensky, A.O.; Tomilin, N.V.; Zalenskaya, I.A.; Teplitz, R.L.; Bradbury, E.M. Telomere-telomere interactions and candidate telomere binding protein(s) in mammalian sperm cells. *Exp. Cell Res.* **1997**, *232*, 29–41. [CrossRef] [PubMed]

57. Greaves, I.K.; Rens, W.; Ferguson-Smith, M.A.; Griffin, D.; Marshall Graves, J.A. Conservation of chromosome arrangement and position of the X in mammalian sperm suggests functional significance. *Chromosom. Res. Int. J. Mol. Supramol. Evol. Asp. Chromosom. Biol.* **2003**, *11*, 503–512. [CrossRef]

58. Watson, J.M.; Meyne, J.; Graves, J.A. Ordered tandem arrangement of chromosomes in the sperm heads of monotreme mammals. *Proc. Natl. Acad. Sci. USA* **1996**, *93*, 10200–10205. [CrossRef] [PubMed]

59. Joffe, B.I.; Solovei, I.V.; Macgregor, H.C. Ordered arrangement and rearrangement of chromosomes during spermatogenesis in two species of planarians (Plathelminthes). *Chromosoma* **1998**, *107*, 173–183. [CrossRef] [PubMed]

60. Solovei, I.V.; Joffe, B.I.; Hori, T.; Thomson, P.; Mizuno, S.; Macgregor, H.C. Unordered arrangement of chromosomes in the nuclei of chicken spermatozoa. *Chromosoma* **1998**, *107*, 184–188. [CrossRef] [PubMed]

61. Li, Y.; Lalancette, C.; Miller, D.; Krawetz, S.A. Characterization of nucleohistone and nucleoprotamine components in the mature human sperm nucleus. *Asian J. Androl.* **2008**, *10*, 535–541. [CrossRef] [PubMed]

62. Nelson, J.E.; Krawetz, S.A. Computer assisted promoter analysis of a human sperm specific nucleoprotein gene cluster. *DNA Seq. J. DNA Seq. Mapp.* **1995**, *5*, 329–337. [CrossRef]

63. Thilagavathi, J.; Venkatesh, S.; Dada, R. Telomere length in reproduction. *Andrologia* **2013**, *45*, 289–304. [CrossRef] [PubMed]

64. Ioannou, D.; Millan, N.M.; Jordan, E.; Tempest, H.G. A new model of sperm nuclear architecture following assessment of the organization of centromeres and telomeres in three-dimensions. *Sci. Rep.* **2017**, *7*, 41585. [CrossRef] [PubMed]

65. Sotolongo, B.; Huang, T.T.F.; Isenberger, E.; Ward, W.S. An endogenous nuclease in hamster, mouse, and human spermatozoa cleaves DNA into loop-sized fragments. *J. Androl.* **2005**, *26*, 272–280. [CrossRef] [PubMed]

66. Wykes, S.M.; Krawetz, S.A. The structural organization of sperm chromatin. *J. Biol. Chem.* **2003**, *278*, 29471–29477. [CrossRef] [PubMed]
67. Ioannou, D.; Tempest, H.G. Does genome organization matter in spermatozoa? A refined hypothesis to awaken the silent vessel. *Syst. Biol. Reprod. Med.* **2018**. [CrossRef] [PubMed]
68. Wiland, E.; Fraczek, M.; Olszewska, M.; Kurpisz, M. Topology of chromosome centromeres in human sperm nuclei with high levels of DNA damage. *Sci. Rep.* **2016**, *6*, 31614. [CrossRef] [PubMed]
69. Noblanc, A.; Peltier, M.; Damon-Soubeyrand, C.; Kerchkove, N.; Chabory, E.; Vernet, P.; Saez, F.; Cadet, R.; Janny, L.; Pons-Rejraji, H.; et al. Epididymis response partly compensates for spermatozoa oxidative defects in snGPx4 and GPx5 double mutant mice. *PLoS ONE* **2012**, *7*, e38565. [CrossRef]

Review

Genetic Instability and Chromatin Remodeling in Spermatids

Tiphanie Cavé, Rebecka Desmarais, Chloé Lacombe-Burgoyne and Guylain Boissonneault *

Department of Biochemistry, Faculty of Medicine and Health Sciences, Université de Sherbrooke, Sherbrooke, QC J1E4K8, Canada; tiphanie.cave@usherbrooke.ca (T.C.); rebecka.desmarais@usherbrooke.ca (R.D.); chloe.lacombe-burgoyne@usherbrooke.ca (C.L.-B.)

* Correspondence: guylain.boissonneault@usherbrooke.ca; Tel.: +1-819-821-8000 (ext. 75443)

Received: 27 December 2018; Accepted: 8 January 2019; Published: 14 January 2019

Abstract: The near complete replacement of somatic chromatin in spermatids is, perhaps, the most striking nuclear event known to the eukaryotic domain. The process is far from being fully understood, but research has nevertheless unraveled its complexity as an expression of histone variants and post-translational modifications that must be finely orchestrated to promote the DNA topological change and compaction provided by the deposition of protamines. That this major transition may not be genetically inert came from early observations that transient DNA strand breaks were detected in situ at chromatin remodeling steps. The potential for genetic instability was later emphasized by our demonstration that a significant number of DNA double-strand breaks (DSBs) are formed and then repaired in the haploid context of spermatids. The detection of DNA breaks by 3′OH end labeling in the whole population of spermatids suggests that a reversible enzymatic process is involved, which differs from canonical apoptosis. We have set the stage for a better characterization of the genetic impact of this transition by showing that post-meiotic DNA fragmentation is conserved from human to yeast, and by providing tools for the initial mapping of the genome-wide DSB distribution in the mouse model. Hence, the molecular mechanism of post-meiotic DSB formation and repair in spermatids may prove to be a significant component of the well-known male mutation bias. Based on our recent observations and a survey of the literature, we propose that the chromatin remodeling in spermatids offers a proper context for the induction of de novo polymorphism and structural variations that can be transmitted to the next generation.

Keywords: spermiogenesis; chromatin remodeling; DNA double-strand breaks; genetic instability; mutations

---

1. Introduction

As one can appreciate from this Special Issue, the proper packaging of the male haploid genome involves finely-regulated molecular events, resulting in a near complete replacement of somatic chromatin and the formation of a highly condensed nucleus. Although the final protamine deposition (protamination) yields a genetically and mechanically stable nucleus, it became rather intuitive that the previous chromatin remodeling steps and resulting change in DNA topology [1] entailed potential genetic hazards. Early observations that H4 hyperacetylation occurs in murine spermatids [2,3] provided the first evidence that chromatin most likely undergoes a transient state of increased sensitivity to endonucleases during this process [4]. H4 hyperacetylation also sets the stage for the bromodomain, testis-specific, protein (Brdt)-mediated histone eviction [5,6]. In situ detection of 3′OH ends in mouse and human spermatids confirmed that H4 hyperacetylation is indeed coincidental, or slightly precedes the formation of the DNA strand breaks that were observed in the whole spermatid population [7]. Because of its potentially significant transgenerational impact, establishing the genetic consequences of this structural transition and the formation of transient DNA strand breaks has been the ultimate objective of our investigation over the past 15 years.

2. DNA Double-Strand Breaks (DSBs) are Intrinsic to the Differentiation Program of Spermatids

Our initial observation that transient DNA strand breaks were observed in the whole population of mouse and human spermatids ruled out that a canonical apoptotic process was involved, as discussed below. Evidence of a similar surge in DNA strand breakage was also reported in rats [8], drosophila [9], grasshoppers (*Eyprepocnemis plorans*) [10] and in algae (*Chara vulgaris*) [11]. Several methods were used to confirm that the transient DNA strand breakage in spermatids also included a significant proportion of DNA double-strand breaks (DSBs). These included neutral comet assay, pulse-field gel electrophoresis [12], γH2AX labelling [13] and qTUNEL assay, whereby double-strand breaks were specifically labelled in solution following a prior step involving DNA nicks and gap filling [14]. Indirect evidence of a DSB repair response based on γH2AX expression must however be taken with caution, as the latter has also been associated with chromatin alteration [15]. Direct methods were also used by our group to show that transient post-meiotic DSBs also form in the fission yeast, lending strong support to the highly conserved nature of this mechanism [16]. Despite their transient character, the formation of DSBs in the haploid context of spermatids represents a genetic threat, because repair must rely solely on end joining processes, as outlined below [17,18]. Considering the lower DNA repair activity reported in condensing spermatids, potential misprocessing of these DSBs would be expected to further increase genetic instability.

3. Potential Mechanism for DSBs Formation and Repair

Chromatin structure is a key determinant of the meiotic DSB landscape [19] and hotspot specification in meiosis, where it has been shown to arise from a combination of factors acting upon histones, including the histone methyltransferase PRDM9 [20,21]. In the mature sperm, protamine affords similar but periodic loop-sized protection against the combined activity of an endogenous nuclease interacting with the nuclear matrix-associated topoisomerase IIB (TOP2B) [22]. Similarly, the dynamic character of the chromatin structure transition in spermatids must dictate the genomic distribution of the DSBs during the differentiation steps, and it stands to reason that DNA strand break hotspots regions should be found, although the endonucleases involved have yet to be identified. The potential involvement of TOP2B in the transient formation of DSB has been inferred from the synchronous detection of TOP2B and the tyrosyl-DNA phosphodiesterase 1 (TDP1), which are known to resolve topoisomerase-mediated DNA damage [23]. Such DNA breaks may have been created from the simple hindrance of the TOP2B catalytic cycle during the nuclear condensation process in elongating spermatids. Using RNA interference, the requirement for TOP2 activity in the post-meiotic DSB formation has been recently demonstrated in the ciliate *Terahymena thermophila* [24]. The extent of DSB formation in elongating spermatids is yet unknown. As previously proposed [25], TOP2B is expected to relax all free supercoils generated from nucleosome eviction, reducing the DNA linking number (Lk) in steps of two per catalytic cycle. Although a great number of DSBs should be generated in this case, the DNA ends remain concealed, and it may not elicit a DNA damage response. This is in sharp contrast to the action of an endonuclease that would relax larger domains of unconstrained supercoils from simple strand breakage due to its inability to relegate DNA ends. However, as suggested above in studies concerning mature sperm, a controlled fragmentation process involving the combined action of TOP2 activity and endonucleases in differentiating spermatids should also be considered. For instance, such an interaction between TOP2A and endonuclease G has been shown to be involved in caspase-independent apoptotic DNA fragmentation, since the mitochondrial endonuclease G is translocated to the nuclei during apoptosis [26,27]. Without leading to cell death, it has been proposed that the apoptotic machinery could be borrowed for various differentiation processes, including spermiogenesis [28,29]. Multiple caspases were found to be expressed in the residual bodies of the Drosophila spermatids to remove the unneeded cytoplasmic content during the process of individualization, but caspases are seemingly kept away from the nucleus [28]. In accordance with an early report from Smith and Haaf [30], our group has found no evidence of complete canonical nuclear apoptosis in the nuclei of mouse spermatids in agreement with the transient character of the DNA strand break formation. Since we observed nuclear translocation of endonuclease G in the nuclei of elongating spermatids (unpublished data), its potential association with topoisomerase in spermatids is currently under investigation, as this could promote a caspase-independent mechanism that would lead to the observed surge in DNA fragmentation. Such a mechanism could be under the control of PARP1 and PARP2, as they were shown to strongly inhibit TOP2B activity of spermatids both in vivo and in vitro [31]. One major hallmark of apoptosis is the resulting mitochondrial outer membrane permeabilization, which releases endonuclease G and apoptosis-inducing factors (AIF) [32–35]. Importantly, although the expression of pro-apoptotic factors has not yet been reported in spermatids, mitochondria are nevertheless known to undergo major structural changes during spermiogenesis. While part of them move to the growing flagellum, other starts to aggregate, and are eventually eliminated by Sertoli cells via phagocytosis or autolysis [36]. It is therefore possible that mitochondrial endonuclease G is released during the autolytic destruction of the outer membrane during the chromatin-remodeling steps. It is worth nothing that even limited, regulated mitochondrial permeabilization can produce DNA damage and genomic instability, without leading to cell death [37], supporting the concept that mitochondrial damage could entail controlled yet reversible DNA fragmentation.

In recent studies, the reversible character of such genome-scale apoptotic-like DNA fragmentation (reversal of apoptosis) has been observed in many instances upon the withdrawal of inducers [38,39]. Striking examples of this are the recovery from global DNA fragmentation observed in the African midge (*P. Vanderplanki*) larva after extreme dehydration [40], similar to the extremotolerant tardigrade species (*R. varieornatus*) [41]. The latter has been shown to express a newly identified, highly basic DNA binding protein (Dsup). The activity of Dsup has been shown to protect transfected cells against radiation and ROS-induced DNA breaks, which is somewhat reminiscent of the protective effect against UV-induced DNA damage that we previously reported for transition proteins [42]. Thus, such a recovery from massive DNA fragmentation, coined "anastasis", indicates that related mechanisms may be operating in spermatids for global DNA repair. However, recovery from apoptosis-like processes can promote mutagenesis and even oncogenic transformation [37], often displaying micronuclei and chromosomal abnormalities [39,43].

Taken together, this compelling evidence suggests that the chromatin remodeling in haploid spermatids, which precludes the use of homologous recombination for templated DSB repair, should create genetic instability [18]. End-joining processes for DSB repair that are likely to operate in a haploid context include single-strand annealing (SSA), microhomology-mediated end joining (MMEJ) or canonical nonhomologous end joining (NHEJ). SSA occurs within repeated sequences and is known to be intrinsically mutagenic [44], whereas, in the absence of canonical NHEJ factors, MMEJ can process the resected DNA ends, using as little as 1-2bp homology when stabilized by PARP [45]. Canonical NHEJ can proceed without sequence homology, and results in insertion, deletion or even chromosomal rearrangement [17]. Whereas meiosis may have evolved mechanisms to prevent these error-prone end-joining processes [18,46], haploid spermatids likely cannot avoid such mutagenic repair mechanisms. However, our initial mapping data indicated that the transient post-meiotic DSBs arise preferentially within repeated elements of the genome, which should minimize the genetic threat associated with the DSB formation [12]. It is important that coding sequences be protected from the global DNA fragmentation process, especially because of the general loss in DNA repair capacity that has been observed as the chromatin remodeling in spermatids proceeds to the final steps [47–49]. In pathological conditions, further alteration in the repair capacity of spermatids could lead to a persistence of DSBs in spermatozoa. Unrepaired DSBs in sperm could be of a lesser concern, given the reported efficient DNA repair activity of the oocytes [50].

4. First Evidence of Genetic Instability in Spermatids

Trinucleotide repeats (TNRs) are the most unstable DNA sequences, and transgenerational expansion beyond a given threshold has been linked to inherited neuromuscular and neurological disorders in offspring [51,52]. Thus, variations in TNR represent an ideal sentinel to monitor genetic instability as a result of faulty DNA repair or chromatin remodeling. Using a transgenic mouse model, the TNR expansion of a CAG repeat within exon 1 of the human *HD* gene was shown to be limited to post-meiotic events in males, and thus does not involve mitotic replication or homologous recombination between chromosomes [53]. Using this mouse model [54], the purification of spermatids into four distinct populations allowed us to further demonstrate that an increased frequency of longer DNA repeat length occurs just following chromatin remodeling, as was observed in mouse step 15–16 spermatids, which is equivalent to the transition between steps 3 and 4 in human spermiogenesis [55]. Interestingly, based on the increased intensity of the individual repeat length, we estimated that approximately 20% of spermatids displayed a shift to a longer repeat length [56]. In vitro experiments suggested that the increase in free superhelical density that must prevail during histone eviction resulted in expansion at a stabilized hairpin [57]. DNA secondary structures are indeed causative factors of expansion [52], and the free supercoils in spermatids offer an ideal context for hairpin extrusion and stabilization of other alternative non-B DNA conformations at repeated elements, including cruciform and left-handed Z-DNA [58,59]. Because non-B DNA structures are preferential substrates for endonucleolytic incisions [60,61], the striking enrichment of DSBs that we observed at repeated elements of the spermatid's genome may therefore not be surprising. Thus, monitoring TNRs length variation during the chromatin remodeling provided the first experimental evidence that genetic instability is an important feature of differentiating spermatids.

5. DNA Fragmentation in Spermatids and the Male Mutation Bias

Over the past few years, next generation sequencing (NGS) of parent–offspring trios has confirmed the clear male bias for the transmission of de novo mutations. Male-biased mutations include single-nucleotide variants, small insertions–deletions (indels), and structural variations [62–67]. Not surprisingly, the greater number of replication cycles in spermatogenesis was originally suspected as the leading cause of de novo mutations, because the male-to-female mutation rate ratio correlates with the male-to-female ratio of the number of cell divisions [68].

The relatively recent observation that transient DNA double-strand breaks are part of the differentiation program of spermatids (and the even more recent results on the genome-wide distribution of these breaks) could explain why only a few reports have considered this process in the etiology of the male mutation bias. Several lines of evidence, however, suggest that the formation of DSBs and repair in the haploid context of spermatids are compatible with the recent NGS data. First, the amount of mutations generated during DNA replication is much lower than the polymerase error rate. Hence, the reliability and extent of DNA repair (or DNA repair rate) becomes prominent in the determination of transmittable de novo mutations [69]. Previous reports confirmed the decline in the general repair capacity during spermiogenesis [49], and the limited response to DSBs in elongating spermatids compared to pre-meiotic cells [47,48,70]. Second, our initial screening in mice indicated that a large part of the transient DSBs in spermatids map to repeated elements of the genome, in accordance with the relative abundance of these intergenic regions. Interestingly, however, DSBs arise at a greater frequency in LINEs and microsatellites relative to their normal representation in the genome [12]. Coincidentally, a higher frequency of de novo mutations was found to arise within repetitive DNA sequences [67], and a strong paternal bias has been reported for mutations within microsatellite repeats [71]. Interestingly, we found the density of DSBs to be four times higher in the Y chromosomes than autosomes, which is compatible with the higher mutability reported for the Y chromosome [72]. Third, and as outlined above, DSBs and end-joining repair processes in haploid spermatids are likely to offer a proper context for male-driven rearrangement. De novo indels and structural variations such as retrotransposon insertions and interchromosomal events were shown to arise preferentially in the paternal germline [66,73], and a similar paternal bias has been reported for copy number variation (CNV) [74,75]. Although replication-based mechanisms could still be responsible for these mutational and structural modifications, they can also arise from the formation of DSBs in the haploid context of spermatids. For instance, in haploid cells, CNV may result from NHEJ, but can also be generated by non-allelic homologous recombination (NAHR) [76], since NAHR is produced by the alignment and subsequent crossover between nonallelic DNA sequence repeats sharing a homology. Both mechanisms require the formation of a DSB [77]. Fourth, it was established that these male-driven variations are generally associated with the transmission of neurodevelopmental disorders [62,63,65,74,78]. For instance, 88% of de novo indels arise on the paternal chromosome and are associated with autism spectrum disorders [65]. When considering only the gene subset, our preliminary gene ontology term analysis showed that DSBs in spermatids arise preferentially within synaptic genes, and with high significance [12]. Hence, the transient DSB formation in spermatids deserves more attention as a potentially significant mechanism by which paternal variation may be transmitted with implications for neurodevelopment.

The transient DSB surge in spermatids may be viewed as a serious threat to the genetic integrity of the differentiating spermatids before it is eventually used for fertilization. However, one must consider that the DSBs are detected in the whole spermatids population and represent the full repertoire of potentially unstable loci seen in a pool of several million cells. The whole genome capture of these DSBs provides a map of their distribution in the whole cell population, if they can be detected by being present in much more than a single cell. A strong DSB hotspot leading to a deleterious mutation or structural variation must therefore be present in a significant subset of spermatids in order to increase the chance of these being selected for fertilization. Pathological conditions that would increase global DNA fragmentation in spermatids could lead to a concomitant increase in the frequency of a given hotspot among cells, thus increasing the chance of that allele being transmitted to the offspring. On the other hand, if a mutational DSBs hotspot is present at a much lower frequency, for instance at 0.01% or lower, then the mutated allele would stand a maximum of 10^{-4} chance of being transmitted for each oocyte being fertilized. In such a case, a beneficial or deleterious mutation would be passed on to the next generation over a much longer (or evolutionary) timescale, provided that a hotspot locus was maintained over time in each species. Interestingly, we observed that selected hotspots are shared between two mouse strains (C57BL/6 vs CD1), suggesting that they can be maintained at least over the evolutionary distance between these inbred and outbred strains. A higher frequency (or density) of such weaker hotspots in a chromosome would confer a faster evolutionary global mutability on an evolutionary timescale, as observed for the Y chromosome.

6. DSBs in Lower Eukaryotes

The apparent conservation of transient DSBs in spermatids between mammals prompted us to investigate whether post-meiotic DSBs may be conserved through the eukaryotic domain, and also be associated with gamete formation in yeast (sporulation) [79]. Being more similar to metazoans, fission yeast (*S. pombe*) represents a better alternative than budding yeasts (*S. cerevisiae*) [80]. In addition, synchronous meiosis can be achieved with the *S. pombe pat1-114* mutant, in which a temperature-sensitive Pat1 (Ran1) protein kinase inhibits meiosis by negatively regulating an RNA-binding protein that controls entry into the meiotic S phase, Mei2. Synchronous meiosis can therefore be induced in a timely and predictable fashion by shifting nitrogen-starved cultures from permissive (25 °C) to restrictive (34 °C) temperatures [81]. One striking observation is that synchronized *S. pombe* displayed a similar post-meiotic surge in DSBs in the absence of apoptosis, when meiosis is induced in the *pat1-114* mutant, or even in the wild-type *FY435/FY436* strain (azygotic meiosis) [16], suggesting that sporulation, much like spermiogenesis, may display a window of genetic instability. Our conclusion is that transient post-meiotic DSBs may be intrinsic to the gamete differentiation program throughout the eukaryotic domain. As outlined above, a major support for this conclusion recently came from the demonstration that DSBs also arise during post-meiotic steps in the ciliate *Tetrahymena thermophila* [24]. These observations point to the discovery of a highly conserved, physiological mechanism that deserves further investigation regarding its genetic impact and evolutionary consequences. Simple eukaryotic models such as yeast offer the possibility of functional genetics analyses, to identify the endonuclease(s) responsible for the transient DSB formation, and eventually determine their impact on adaptation and evolution over several generations. In-gel nuclease assays in the synchronized pat1-114 mutant have already led to the identification of a candidate mitochondrial endonuclease (Pnu1), an homolog of the *S. cerevisiae* Nuc1p that has been described as part of the caspase-independent apoptotic pathway [82]. Interestingly, the mammalian homolog of Nuc1p is the mitochondrial endonuclease G (Endo G) discussed above, which is also involved in caspase-independent apoptosis. These early observations in yeast lend support to the proposal that the transient post-meiotic DSBs may indeed borrow components of the apoptotic machinery in a controlled, reversible manner.

7. Conclusions and Future Directions

Given the non-templated DNA repair in haploid spermatids, transient DSB formation may represent an important component of the male mutation bias and the etiology of neurological disorders, adding to the genetic variation provided by meiosis. Repair heterogeneity at these potential hotspots would produce a repertoire of genetic polymorphisms, given the large population of spermatozoa produced over time. In addition to the chromosome reshuffling provided by meiosis, each offspring would also inherit a given set of mutations created by the chromatin remodeling in the spermatid of origin. Because synaptic genes were found to be specifically targeted by DSBs and should therefore harbor more mutational hotspots, pathological conditions (or aging) leading to a global rise in DSB formation would then increase the odds of transmitting de novo variation in neurodevelopmental genes. This hypothesis is thus in agreement with the reported correlation between the father's age at conception and the risk of transmitting neurological disorders. Further investigation should therefore be aimed at deciphering whether mutational DSBs hotspots arise within neurodevelopmental genes and how they are altered under pathological conditions, or following exposure to xenobiotics. Monitoring the distribution and number of DSBs in elongating spermatids should be clearly emphasized, as they represent a much higher threat compared to single-strand breaks. Despite the reported DNA repair capacity of the oocyte, the transgenerational consequences of an increased number of persistent DSBs in sperm deserves some attention for future investigations.

Author Contributions: Conceptualization, T.C. and G.B.; investigation, T.C., R.D., C.L.-B., G.G.; writing—original draft preparation G.B.; writing—review and editing, T.C., R.D., C.L.-B., G.B.

Funding: This work was supported by a grant from the Canadian Institutes of Health Research (#MOP-136925) to G.B.

Conflicts of Interest: The authors declare no conflict of interest. The funders had no role in the design of the study; in the collection, analyses, or interpretation of data; in the writing of the manuscript, or in the decision to publish the results.

References

1. Risley, M.S.; Einheber, S.; Bumcrot, D.A. Changes in DNA topology during spermatogenesis. *Chromosoma* **1986**, *94*, 217–227. [CrossRef] [PubMed]
2. Grimes, S.R., Jr.; Henderson, N. Hyperacetylation of histone H4 in rat testis spermatids. *Exp. Cell Res.* **1984**, *152*, 91–97. [CrossRef]
3. Meistrich, M.L.; Trostle-Weige, P.K.; Lin, R.; Bhatnagar, Y.M.; Allis, C.D. Highly acetylated H4 is associated with histone displacement in rat spermatids. *Mol. Reprod. Dev.* **1992**, *31*, 170–181. [CrossRef] [PubMed]
4. Martinez-Lopez, W.; Folle, G.A.; Obe, G.; Jeppesen, P. Chromosome regions enriched in hyperacetylated histone H4 are preferred sites for endonuclease- and radiation-induced breakpoints. *Chromosome Res.* **2001**, *9*, 69–75. [CrossRef] [PubMed]

5. Goudarzi, A.; Shiota, H.; Rousseaux, S.; Khochbin, S. Genome-scale acetylation-dependent histone eviction during spermatogenesis. *J. Mol. Biol.* **2014**, *426*, 3342–3349. [CrossRef] [PubMed]

6. Shiota, H.; Barral, S.; Buchou, T.; Tan, M.; Coute, Y.; Charbonnier, G.; Reynoird, N.; Boussouar, F.; Gerard, M.; Zhu, M.; et al. Nut Directs p300-Dependent, Genome-Wide H4 Hyperacetylation in Male Germ Cells. *Cell Rep.* **2018**, *24*, 3477–3487. [CrossRef]

7. Marcon, L.; Boissonneault, G. Transient DNA strand breaks during mouse and human spermiogenesis new insights in stage specificity and link to chromatin remodeling. *Biol. Reprod.* **2004**, *70*, 910–918. [CrossRef]

8. Meyer-Ficca, M.L.; Scherthan, H.; Bürkle, A.; Meyer, R.G. Poly(ADP-ribosyl)ation during chromatin remodeling steps in rat spermiogenesis. *Chromosoma* **2005**, *114*, 67–74. [CrossRef]

9. Rathke, C.; Baarends, W.M.; Jayaramaiah-Raja, S.; Bartkuhn, M.; Renkawitz, R.; Renkawitz-Pohl, R. Transition from a nucleosome-based to a protamine-based chromatin configuration during spermiogenesis in *Drosophila*. *J. Cell Sci.* **2007**, *120*, 1689–1700. [CrossRef]

10. Cabrero, J.; Palomino-Morales, R.J.; Camacho, J.P.M. The DNA-repair Ku70 protein is located in the nucleus and tail of elongating spermatids in grasshoppers. *Chromosome Res. Int. J. Mol. Supramol. Evol. Asp. Chromosome Biol.* **2007**, *15*, 1093–1100. [CrossRef]

11. Wojtczak, A.; Popłońska, K.; Kwiatkowska, M. Phosphorylation of H2AX histone as indirect evidence for double-stranded DNA breaks related to the exchange of nuclear proteins and chromatin remodeling in *Chara vulgaris* spermiogenesis. *Protoplasma* **2008**, *233*, 263–267. [CrossRef] [PubMed]

12. Gregoire, M.C.; Leduc, F.; Morin, M.H.; Cave, T.; Arguin, M.; Richter, M.; Jacques, P.E.; Boissonneault, G. The DNA double-strand "breakome" of mouse spermatids. *Cell Mol. Life Sci.* **2018**, *75*, 2859–2872. [CrossRef] [PubMed]

13. Leduc, F.; Maquennehan, V.; Nkoma, G.B.; Boissonneault, G. DNA damage response during chromatin remodeling in elongating spermatids of mice. *Biol. Reprod.* **2008**, *78*, 324–332. [CrossRef] [PubMed]

14. Gregoire, M.C.; Massonneau, J.; Leduc, F.; Arguin, M.; Brazeau, M.A.; Boissonneault, G. Quantification and genome-wide mapping of DNA double-strand breaks. *DNA Repair* **2016**, *48*, 63–68. [CrossRef] [PubMed]

15. Turinetto, V.; Giachino, C. Multiple facets of histone variant H2AX: A DNA double-strand-break marker with several biological functions. *Nucleic Acids Res.* **2015**, *43*, 2489–2498. [CrossRef] [PubMed]

16. Cave, T.; Gregoire, M.C.; Brazeau, M.A.; Boissonneault, G. Post-meiotic DNA double-strand breaks are conserved in fission yeast. *Int. J. Biochem. Cell Biol.* **2018**. [CrossRef] [PubMed]

17. Rodgers, K.; McVey, M. Error-Prone Repair of DNA Double-Strand Breaks. *J. Cell Physiol.* **2016**, *231*, 15–24. [CrossRef] [PubMed]

18. Kim, S.; Peterson, S.E.; Jasin, M.; Keeney, S. Mechanisms of germ line genome instability. *Semin. Cell Dev. Biol.* **2016**, *54*, 177–187. [CrossRef]

19. Pan, J.; Sasaki, M.; Kniewel, R.; Murakami, H.; Blitzblau, H.G.; Tischfield, S.E.; Zhu, X.; Neale, M.J.; Jasin, M.; Socci, N.D.; et al. A hierarchical combination of factors shapes the genome-wide topography of yeast meiotic recombination initiation. *Cell* **2011**, *144*, 719–731. [CrossRef]

20. Lange, J.; Yamada, S.; Tischfield, S.E.; Pan, J.; Kim, S.; Zhu, X.; Socci, N.D.; Jasin, M.; Keeney, S. The Landscape of Mouse Meiotic Double-Strand Break Formation, Processing, and Repair. *Cell* **2016**, *167*, 695–708. [CrossRef]

21. Baudat, F.; Buard, J.; Grey, C.; Fledel-Alon, A.; Ober, C.; Przeworski, M.; Coop, G.; de Massy, B. PRDM9 is a major determinant of meiotic recombination hotspots in humans and mice. *Science* **2010**, *327*, 836–840. [CrossRef] [PubMed]

22. Shaman, J.A.; Prisztoka, R.; Ward, W.S. Topoisomerase IIB and an extracellular nuclease interact to digest sperm DNA in an apoptotic-like manner. *Biol. Reprod.* **2006**, *75*, 741–748. [CrossRef] [PubMed]

23. Murai, J.; Huang, S.Y.; Das, B.B.; Dexheimer, T.S.; Takeda, S.; Pommier, Y. Tyrosyl-DNA phosphodiesterase 1 (TDP1) repairs DNA damage induced by topoisomerases I and II and base alkylation in vertebrate cells. *J. Biol. Chem.* **2012**, *287*, 12848–12857. [CrossRef] [PubMed]

24. Akematsu, T.; Fukuda, Y.; Garg, J.; Fillingham, J.S.; Pearlman, R.E.; Loidl, J. Post-meiotic DNA double-strand breaks occur in *Tetrahymena*, and require Topoisomerase II and Spo11. *Elife* **2017**, *6*. [CrossRef] [PubMed]

25. Ward, W.S. Regulating DNA Supercoiling: Sperm Points the Way. *Biol. Reprod.* **2011**, *84*, 841–843. [CrossRef] [PubMed]

26. Solovyan, V.T. Characterization of apoptotic pathway associated with caspase-independent excision of DNA loop domains. *Exp. Cell Res.* **2007**, *313*, 1347–1360. [CrossRef] [PubMed]

27. Varecha, M.; Potesilova, M.; Matula, P.; Kozubek, M. Endonuclease G interacts with histone H2B and DNA topoisomerase II α during apoptosis. *Mol. Cell Biochem.* **2012**, *363*, 301–307. [CrossRef]

28. Arama, E.; Agapite, J.; Steller, H. Caspase activity and a specific cytochrome C are required for sperm differentiation in *Drosophila*. *Dev. Cell* **2003**, *4*, 687–697. [CrossRef]

29. Cagan, R.L. Spermatogenesis: Borrowing the apoptotic machinery. *Curr. Biol.* **2003**, *13*, R600–R602. [CrossRef]

30. Smith, A.; Haaf, T. DNA nicks and increased sensitivity of DNA to fluorescence in situ end labeling during functional spermiogenesis. *Biotechniques* **1998**, *25*, 496–502. [CrossRef]

31. Meyer-Ficca, M.L.; Lonchar, J.D.; Ihara, M.; Meistrich, M.L.; Austin, C.A.; Meyer, R.G. Poly(ADP-Ribose) Polymerases PARP1 and PARP2 Modulate Topoisomerase II Beta (TOP2B) Function during Chromatin Condensation in Mouse Spermiogenesis. *Biol. Reprod.* **2011**, *84*, 900–909. [CrossRef] [PubMed]

32. Cande, C.; Vahsen, N.; Garrido, C.; Kroemer, G. Apoptosis-inducing factor (AIF): Caspase-independent after all. *Cell Death Differ* **2004**, *11*, 591–595. [CrossRef] [PubMed]

33. Zamzami, N.; Susin, S.A.; Marchetti, P.; Hirsch, T.; Gómez-Monterrey, I.; Castedo, M.; Kroemer, G. Mitochondrial control of nuclear apoptosis. *J. Exp. Med.* **1996**, *183*, 1533–1544. [CrossRef] [PubMed]

34. Li, L.Y.; Luo, X.; Wang, X. Endonuclease G is an apoptotic DNase when released from mitochondria. *Nature* **2001**, *412*, 95–99. [CrossRef] [PubMed]

35. Joza, N.; Susin, S.A.; Daugas, E.; Stanford, W.L.; Cho, S.K.; Li, C.Y.; Sasaki, T.; Elia, A.J.; Cheng, H.Y.; Ravagnan, L.; et al. Essential role of the mitochondrial apoptosis-inducing factor in programmed cell death. *Nature* **2001**, *410*, 549–554. [CrossRef] [PubMed]

36. Sun, X.; Yang, W.-X. Mitochondria: Transportation, distribution and function during spermiogenesis. *Adv. Biosci. Biotechnol.* **2010**, *1*, 13. [CrossRef]

37. Ichim, G.; Lopez, J.; Ahmed, S.U.; Muthalagu, N.; Giampazolias, E.; Delgado, M.E.; Haller, M.; Riley, J.S.; Mason, S.M.; Athineos, D.; et al. Limited mitochondrial permeabilization causes DNA damage and genomic instability in the absence of cell death. *Mol. Cell* **2015**, *57*, 860–872. [CrossRef]

38. Sun, G.; Guzman, E.; Balasanyan, V.; Conner, C.M.; Wong, K.; Zhou, H.R.; Kosik, K.S.; Montell, D.J. A molecular signature for anastasis, recovery from the brink of apoptotic cell death. *J. Cell Biol.* **2017**, *216*, 3355–3368. [CrossRef]

39. Tang, H.L.; Tang, H.M.; Mak, K.H.; Hu, S.; Wang, S.S.; Wong, K.M.; Wong, C.S.; Wu, H.Y.; Law, H.T.; Liu, K.; et al. Cell survival, DNA damage, and oncogenic transformation after a transient and reversible apoptotic response. *Mol. Biol. Cell* **2012**, *23*, 2240–2252. [CrossRef]

40. Gusev, O.; Suetsugu, Y.; Cornette, R.; Kawashima, T.; Logacheva, M.D.; Kondrashov, A.S.; Penin, A.A.; Hatanaka, R.; Kikuta, S.; Shimura, S.; et al. Comparative genome sequencing reveals genomic signature of extreme desiccation tolerance in the anhydrobiotic midge. *Nat. Commun.* **2014**, *5*, 4784. [CrossRef]

41. Hashimoto, T.; Horikawa, D.D.; Saito, Y.; Kuwahara, H.; Kozuka-Hata, H.; Shin, I.T.; Minakuchi, Y.; Ohishi, K.; Motoyama, A.; Aizu, T.; et al. Extremotolerant tardigrade genome and improved radiotolerance of human cultured cells by tardigrade-unique protein. *Nat. Commun.* **2016**, *7*, 12808. [CrossRef] [PubMed]

42. Caron, N.; Veilleux, S.; Boissonneault, G. Stimulation of DNA Repair by the spermatidal TP1 protein. *Mol. Reprod. Dev.* **2001**, *58*, 437–443. [CrossRef]

43. Tang, H.M.; Talbot, C.C., Jr.; Fung, M.C.; Tang, H.L. Molecular signature of anastasis for reversal of apoptosis. *F1000Res* **2017**, *6*, 43. [CrossRef] [PubMed]

44. Mimitou, E.P.; Symington, L.S. DNA end resection: Many nucleases make light work. *DNA Repair* **2009**, *8*, 983–995. [CrossRef] [PubMed]

45. Audebert, M.; Salles, B.; Calsou, P. Involvement of poly(ADP-ribose) polymerase-1 and XRCC1/DNA ligase III in an alternative route for DNA double-strand breaks rejoining. *J. Biol. Chem.* **2004**, *279*, 55117–55126. [CrossRef] [PubMed]

46. Hochwagen, A.; Amon, A. Checking your breaks: Surveillance mechanisms of meiotic recombination. *Curr. Biol.* **2006**, *16*, R217–R228. [CrossRef] [PubMed]

47. Olsen, A.-K.; Lindeman, B.; Wiger, R.; Duale, N.; Brunborg, G. How do male germ cells handle DNA damage? *Toxicol. Appl. Pharmacol.* **2005**, *207*, 521–531. [CrossRef]

48. Ahmed, E.A.; Scherthan, H.; de Rooij, D.G. DNA Double Strand Break Response and Limited Repair Capacity in Mouse Elongated Spermatids. *Int. J. Mol. Sci.* **2015**, *16*, 29923–29935. [CrossRef]

49. Marchetti, F.; Wyrobek, A.J. DNA Repair decline during mouse spermiogenesis results in the accumulation of heritable DNA damage. *DNA Repair* **2008**, *7*, 572–581. [CrossRef]

50. Stringer, J.M.; Winship, A.; Liew, S.H.; Hutt, K. The capacity of oocytes for DNA Repair. *Cell Mol. Life Sci.* **2018**, *75*, 2777–2792. [CrossRef]

51. Castel, A.L.; Cleary, J.D.; Pearson, C.E. Repeat instability as the basis for human diseases and as a potential target for therapy. *Nat. Rev. Mol. Cell Biol.* **2010**, *11*, 165. [CrossRef] [PubMed]

52. McMurray, C.T. Mechanisms of trinucleotide repeat instability during human development. *Nat. Rev. Genet.* **2010**, *11*, 786–799. [CrossRef] [PubMed]

53. Kovtun, I.V.; McMurray, C.T. Trinucleotide expansion in haploid germ cells by gap repair. *Nat. Genet.* **2001**, *27*, 407–411. [CrossRef] [PubMed]

54. Mangiarini, L.; Sathasivam, K.; Seller, M.; Cozens, B.; Harper, A.; Hetherington, C.; Lawton, M.; Trottier, Y.; Lehrach, H.; Davies, S.W.; et al. Exon 1 of the *HD* gene with an expanded CAG repeat is sufficient to cause a progressive neurological phenotype in transgenic mice. *Cell* **1996**, *87*, 493–506. [CrossRef]

55. Simard, O.; Leduc, F.; Acteau, G.; Arguin, M.; Gregoire, M.C.; Brazeau, M.A.; Marois, I.; Richter, M.V.; Boissonneault, G. Step-specific Sorting of Mouse Spermatids by Flow Cytometry. *J. Vis. Exp.* **2015**, e53379. [CrossRef] [PubMed]

56. Simard, O.; Grégoire, M.-C.; Arguin, M.; Brazeau, M.-A.; Leduc, F.; Marois, I.; Richter, M.V.; Boissonneault, G. Instability of trinucleotidic repeats during chromatin remodeling in spermatids. *Hum. Mutat.* **2014**, *35*, 1280–1284. [CrossRef] [PubMed]

57. Simard, O.; Niavarani, S.R.; Gaudreault, V.; Boissonneault, G. Torsional stress promotes trinucleotidic expansion in spermatids. *Mutat. Res.* **2017**, *800–802*, 1–7. [CrossRef]

58. Brázda, V.; Laister, R.C.; Jagelská, E.B.; Arrowsmith, C. Cruciform structures are a common DNA feature important for regulating biological processes. *BMC Mol. Biol.* **2011**, *12*, 33. [CrossRef]

59. Bacolla, A.; Wells, R.D. Non-B DNA conformations, genomic rearrangements, and human disease. *J. Biol. Chem.* **2004**, *279*, 47411–47414. [CrossRef]

60. Zhang, T.; Huang, J.; Gu, L.; Li, G.-M. In vitro repair of DNA hairpins containing various numbers of CAG/CTG trinucleotide repeats. *DNA Repair* **2012**, *11*, 201–209. [CrossRef]

61. Zhao, J.; Wang, G.; Del Mundo, I.M.; McKinney, J.A.; Lu, X.; Bacolla, A.; Boulware, S.B.; Zhang, C.; Zhang, H.; Ren, P.; et al. Distinct Mechanisms of Nuclease-Directed DNA-Structure-Induced Genetic Instability in Cancer Genomes. *Cell Rep.* **2018**, *22*, 1200–1210. [CrossRef] [PubMed]

62. Kong, A.; Frigge, M.L.; Masson, G.; Besenbacher, S.; Sulem, P.; Magnusson, G.; Gudjonsson, S.A.; Sigurdsson, A.; Jonasdottir, A.; Jonasdottir, A.; et al. Rate of de novo mutations and the importance of father's age to disease risk. *Nature* **2012**, *488*, 471–475. [CrossRef] [PubMed]

63. O'Roak, B.J.; Vives, L.; Girirajan, S.; Karakoc, E.; Krumm, N.; Coe, B.P.; Levy, R.; Ko, A.; Lee, C.; Smith, J.D.; et al. Sporadic autism exomes reveal a highly interconnected protein network of de novo mutations. *Nature* **2012**, *485*, 246–250. [CrossRef] [PubMed]

64. Neale, B.M.; Kou, Y.; Liu, L.; Ma'ayan, A.; Samocha, K.E.; Sabo, A.; Lin, C.-F.; Stevens, C.; Wang, L.-S.; Makarov, V.; et al. Patterns and rates of exonic de novo mutations in autism spectrum disorders. *Nature* **2012**, *485*, 242–245. [CrossRef] [PubMed]

65. Dong, S.; Walker, M.F.; Carriero, N.J.; DiCola, M.; Willsey, A.J.; Ye, A.Y.; Waqar, Z.; Gonzalez, L.E.; Overton, J.D.; Frahm, S.; et al. De novo insertions and deletions of predominantly paternal origin are associated with autism spectrum disorder. *Cell Rep.* **2014**, *9*, 16–23. [CrossRef] [PubMed]

66. Kloosterman, W.P.; Francioli, L.C.; Hormozdiari, F.; Marschall, T.; Hehir-Kwa, J.Y.; Abdellaoui, A.; Lameijer, E.W.; Moed, M.H.; Koval, V.; Renkens, I.; et al. Characteristics of de novo structural changes in the human genome. *Genome Res.* **2015**, *25*, 792–801. [CrossRef]

67. Turner, T.N.; Coe, B.P.; Dickel, D.E.; Hoekzema, K.; Nelson, B.J.; Zody, M.C.; Kronenberg, Z.N.; Hormozdiari, F.; Raja, A.; Pennacchio, L.A.; et al. Genomic Patterns of De Novo Mutation in Simplex Autism. *Cell* **2017**, *171*, 710–722. [CrossRef]

68. Hurst, L.D.; Ellegren, H. Sex biases in the mutation rate. *Trends Genet.* **1998**, *14*, 446–452. [CrossRef]

69. Acuna-Hidalgo, R.; Veltman, J.A.; Hoischen, A. New insights into the generation and role of de novo mutations in health and disease. *Genome Biol.* **2016**, *17*, 241. [CrossRef]

70. Codrington, A.M.; Hales, B.F.; Robaire, B. Spermiogenic germ cell phase-specific DNA damage following cyclophosphamide exposure. *J. Androl.* **2004**, *25*, 354–362. [CrossRef]

71. Sun, J.X.; Helgason, A.; Masson, G.; Ebenesersdottir, S.S.; Li, H.; Mallick, S.; Gnerre, S.; Patterson, N.; Kong, A.; Reich, D.; et al. A direct characterization of human mutation based on microsatellites. *Nat. Genet.* **2012**, *44*, 1161–1165. [CrossRef] [PubMed]

72. Repping, S.; van Daalen, S.K.; Brown, L.G.; Korver, C.M.; Lange, J.; Marszalek, J.D.; Pyntikova, T.; van der Veen, F.; Skaletsky, H.; Page, D.C.; et al. High mutation rates have driven extensive structural polymorphism among human Y chromosomes. *Nat. Genet.* **2006**, *38*, 463–467. [CrossRef] [PubMed]

73. Sayres, M.A.W.; Makova, K.D. Genome analyses substantiate male mutation bias in many species. *Bioessays News Rev. Mol. Cell. Dev. Biol.* **2011**, *33*, 938–945. [CrossRef] [PubMed]

74. Hehir-Kwa, J.Y.; Rodriguez-Santiago, B.; Vissers, L.E.; de Leeuw, N.; Pfundt, R.; Buitelaar, J.K.; Perez-Jurado, L.A.; Veltman, J.A. De novo copy number variants associated with intellectual disability have a paternal origin and age bias. *J. Med. Genet.* **2011**, *48*, 776–778. [CrossRef] [PubMed]

75. Ma, R.; Deng, L.; Xia, Y.; Wei, X.; Cao, Y.; Guo, R.; Zhang, R.; Guo, J.; Liang, D.; Wu, L. A clear bias in parental origin of de novo pathogenic CNVs related to intellectual disability, developmental delay and multiple congenital anomalies. *Sci. Rep.* **2017**, *7*, 44446. [CrossRef] [PubMed]

76. Zhang, H.; Zeidler, A.F.; Song, W.; Puccia, C.M.; Malc, E.; Greenwell, P.W.; Mieczkowski, P.A.; Petes, T.D.; Argueso, J.L. Gene copy-number variation in haploid and diploid strains of the yeast *Saccharomyces cerevisiae*. *Genetics* **2013**, *193*, 785–801. [CrossRef] [PubMed]

77. Zhang, F.; Gu, W.; Hurles, M.E.; Lupski, J.R. Copy number variation in human health, disease, and evolution. *Annu. Rev. Genom. Hum. Genet.* **2009**, *10*, 451–481. [CrossRef] [PubMed]

78. Brandler, W.M.; Antaki, D.; Gujral, M.; Kleiber, M.L.; Whitney, J.; Maile, M.S.; Hong, O.; Chapman, T.R.; Tan, S.; Tandon, P.; et al. Paternally inherited cis-regulatory structural variants are associated with autism. *Science* **2018**, *360*, 327–331. [CrossRef]

79. Govin, J.; Berger, S.L. Genome reprogramming during sporulation. *Int. J. Dev. Biol.* **2009**, *53*, 425–432. [CrossRef]

80. Cipak, L.; Polakova, S.; Hyppa, R.W.; Smith, G.R.; Gregan, J. Synchronized fission yeast meiosis using an ATP analog-sensitive Pat1 protein kinase. *Nat. Protoc.* **2014**, *9*, 223–231. [CrossRef]

81. Bahler, J.; Schuchert, P.; Grimm, C.; Kohli, J. Synchronized meiosis and recombination in fission yeast: Observations with pat1-114 diploid cells. *Curr. Genet.* **1991**, *19*, 445–451. [CrossRef] [PubMed]

82. Buttner, S.; Eisenberg, T.; Carmona-Gutierrez, D.; Ruli, D.; Knauer, H.; Ruckenstuhl, C.; Sigrist, C.; Wissing, S.; Kollroser, M.; Frohlich, K.U.; et al. Endonuclease G regulates budding yeast life and death. *Mol. Cell* **2007**, *25*, 233–246. [CrossRef] [PubMed]

Review

Single and Double Strand Sperm DNA Damage: Different Reproductive Effects on Male Fertility

Jordi Ribas-Maynou * and Jordi Benet *

Unitat de Biologia Cel·lular i Genètica Mèdica, Departament de Biologia Cel·lular, Fisiologia i Immunologia, Facultat de Medicina, Universitat Autònoma de Barcelona, 08193 Bellaterra, Spain
* Correspondence: jordi_ri@hotmail.com (J.R.-M.); jordi.benet@uab.cat (J.B.); Tel.: +34-93-581-1773 (J.B.)

Received: 30 December 2018; Accepted: 29 January 2019; Published: 31 January 2019

Abstract: Reproductive diseases have become a growing worldwide problem and male factor plays an important role in the reproductive diagnosis, prognosis and design of assisted reproductive treatments. Sperm cell holds the mission of carrying the paternal genetic complement to the oocyte in order to contribute to an euploid zygote with proper DNA integrity. Sperm DNA fragmentation had been used for decades as a male fertility test, however, its usefulness have arisen multiple debates, especially around Intracytoplasmic Sperm Injection (ICSI) treatments. In the recent years, it has been described that different types of sperm DNA breaks (single and double strand DNA breaks) cause different clinical reproductive effects. On one hand, single-strand DNA breaks are present extensively as a multiple break points in all regions of the genome, are related to oxidative stress and cause a lack of clinical pregnancy or an increase of the conception time. On the other hand, double-strand DNA breaks are mainly localized and attached to the sperm nuclear matrix as a very few break points, are possibly related to a lack of DNA repair in meiosis and cause a higher risk of miscarriage, low embryo quality and higher risk of implantation failure in ICSI cycles. The present work also reviews different studies that may contribute in the understanding of sperm chromatin as well as treatments to prevent sperm DNA damage.

Keywords: sperm DNA damage; DNA fragmentation; infertility; assisted reproduction; miscarriage; implantation

---

1. Introduction

Different fertility societies around the globe and the World Health Organization estimate that infertility is present in between 7% and 15% of couples in reproductive age [1,2]. In a high number of cases female factors and especially female age [3], are the most important causes of infertility, however, different male factors are present in at least 50% of the couples presenting this disorder [4]. Due to the high percentage of incidence in the pathology, recent research suggests that sperm cell and sperm DNA may have a major influence not only in natural conception but also in fertility treatments [5,6].

In front of a fertility disorder or a fertility treatment, microscopic semen analysis measuring sperm concentration, motility and morphology has been the traditional and important first approach to male infertility and, although a high decrease of these parameters had been associated to a lack of achievement of natural pregnancy [7] and nowadays home-based technologies in order to advance the first diagnosis are emerging [8]. However, in most cases these parameters are not indicative of the positive performance of assisted reproduction techniques (ART) [5,9]. In fact, although they are improving, ICSI treatments reached limited implantation rates [10]. Because of that, a deeper study is necessary in most cases to elucidate the alteration in order to design the best treatment in each case.

2. Sperm DNA and Sperm DNA Damage

Spermatogenesis is a very complex cellular process that implies both meiosis and cell differentiation. The main stage of meiosis is in prophase I where, spermatocytes deliberately produce double-strand DNA breaks (DSB) through Spo11 protein [11,12]. These DSB are necessary for homologous chromosomes to allow DNA recombination. Then, after strand invasion, DSB activate the DNA repair machinery through the protein kinase ataxia-telangiectasia mutated (ATM) in order to repair the free ends and therefore generate the chiasma by homologous recombination and ATM is also responsible of inhibiting the formation of new DSB by Spo11 [12,13]. After meiosis, haploid round spermatids suffer a cell differentiation, loosing most part of their cytoplasm and acquiring midpiece and flagellum in order to possess motility after ejaculation [14]. However, in terms of chromatin, the most important change happening in spermatids is the exchange of histones by protamines, which extraordinarily compact about 85% of the human sperm DNA in toroidal structures tied between them and bond to the nuclear matrix by the matrix attachment regions (MAR regions) (Figure 1). These MAR regions remain compacted by histones and represent a very small part of the genome estimated to be around 15% of the human sperm chromatin [15,16]. This high-grade of DNA compaction with protamines, coupled to a motile architecture of the cell, give the sperm the perfect features to carry male genetic material to oocyte to form the zygote. It is obvious that if this male genetic material contains alterations, these may affect the zygote somehow [17]. In fact, it is undeniable that DNA breaks induce a cellular response in somatic cells leading to an activation of DNA repair machinery, apoptosis or cell transformation, being the basis of cancer and other diseases [18,19]. Different works in embryos analysing the effect of induced DNA breaks in animal sperm cells through radiation observed multiple chromosomal alterations such as chromosome breaks, translocations, fusions and acentric fragments in the zygote [17,20].

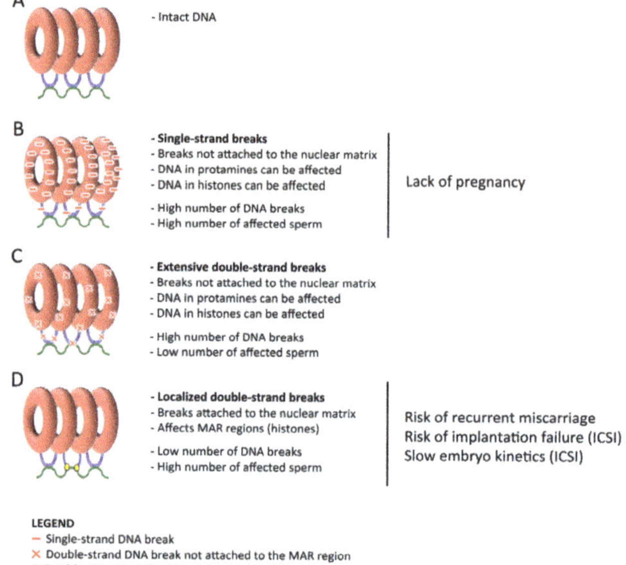

Figure 1. Schematic structure of the sperm DNA compacted in protamines that form toroid structures (red) linked by MAR regions (matrix attachment regions) compacted in histones (blue) and attached to the nuclear matrix (green). (**A**) represents an intact chromatin. (**B**) represents chromatin with single-strand breaks (red lines). (**C**) represents chromatin with extensive double-strand breaks (red cross). (**D**) represents chromatin with localized double-strand breaks attached to the nuclear matrix (yellow circle).

In the last decade, the previous evidences suggested the incorporation of the sperm DNA fragmentation tests as a promising analysis in male reproduction and multiple studies were performed in the field since then [21]. Regarding natural conception, multiple works show a relation of sperm DNA fragmentation (SDF) to a lack of clinical pregnancy and an increase of time of conception [22–24]. However, after ICSI procedures, opposite results were found by different research groups regarding embryo quality, implantation and pregnancy outcomes, being some studies that show a positive relation of SDF [25–28] and others that show a negative relation of SDF to clinical outcomes [29–33]. This controversy, coupled that only a few studies were conducted in a prospective and double blind manner, led the American Society for Reproductive Medicine to refuse its routine use in 2013 [34]. However, some promising results arisen in the last years might be the explanation why the traditionally measured sperm DNA damage present a lack of predictive power in ICSI.

The debate in sperm DNA fragmentation started regarding which of all DNA analysis techniques, that rely on different mechanisms for DNA breaks detection, was the best for the male infertility diagnosis. Understanding the basis of each technique and the correlations between them is critical to understand their implications in the male fertility diagnosis and to compare between them. Techniques are explained in the following part of the review and are summarized in Table 1.

On one hand, the most used techniques for the analysis of sperm DNA fragmentation have traditionally been the Terminal deoxynucleotidyl transferase dUTP nick end labelling (TUNEL), Sperm Chromatin Structure Assay (SCSA) and Sperm Chromatin Dispersion (SCD) test. These techniques offer a unique value of sperm with DNA fragmentation, independently of the type (single and double-strand DNA breaks) and the region (toroids compacted in protamines or MAR regions compacted in histones).

TUNEL assay [35] relies on a terminal TdT transferase for the labelling of 3′ free ends of DNA, resulting in a higher labelling on fragmented sperm cells. Different modifications have been introduced in the protocol in order to increase its sensitivity in sperm cells, such as the use of a previous DNA decompaction using dithiothreitol (DTT) or the use of flow cytometer [36–38].

SCSA is based on an acid denaturation of the chromatin and staining with acridine orange. When DNA breaks are present, chromatin is more susceptible to denaturation and acridine orange accumulates in the DNA emitting in red fluorescence. When DNA breaks are not present, acridine orange intercalates in the double helix and emits in green fluorescence. Fluorescence is captured using a cytometer in order to determine DNA fragmentation [39].

SCD test uses a sperm lysis solution based on DTT, sodium dodecyl sulphate (SDS) and NaCl to remove the sperm membrane and protamines, that causes the formation of DNA haloes, which allow the differentiation of fragmented and non-fragmented sperm cells [40].

Table 1. Techniques for the detection of different types of DNA damage.

Technique	Basic Principle	Advantages	Disadvantages	Type of DNA Damage Detected	Clinical Effect
TUNEL	Labelling of 3′ free ends with a TdT transferase. Breaks are directly labelled.	· Highly standardized protocol.	· Need of flow cytometer for higher number of analysed cells. · Sensitivity for the detection of DNA breaks in sperm cells. · No detection of MAR-region attached DSB.	· Single-strand breaks. · Extensive DSB.	· Pregnancy achievement.
SCSA	Acid denaturation followed by staining with Acridine Orange. DNA with breaks is more susceptible to denaturating.	· Standardized and fast protocol. · Differentiation of immature sperm cells (HDS%).	· Need of flow cytometer. · No detection of MAR-region attached DSB.	· Single-strand breaks. · Extensive DSB.	· Pregnancy achievement.
SCD	Acid denaturation, lysis of sperm membranes and extraction of protamines using detergent and salt. Non-fragmented sperm cells form a halo and fragmented sperm cells do not form halo (form a huge halo that cannot be seen at the optic microscope)	· Highly standardized protocol.	· Non-standardized analysis. · Number of analysed sperm cells · No detection of MAR-region attached DSB.	· Single-strand breaks. · Extensive DSB.	· Pregnancy achievement.
Alkaline Comet	Lysis of sperm membranes and extraction of protamines, alkaline denaturation and electrophoresis at alkaline pH. DNA breaks migrate towards cathode forming a DNA tail.	· Differentiation of mostly single strand DNA breaks at 4 minutes of electrophoresis. · Modulation: longer electrophoresis time may allow elucidating total DNA damage. · Allow quantification of DNA breaks with specific software.	· Technique and analysis are not standardized between laboratories. · No detection of MAR-region DSB. · Studies comparing different electrophoresis times are needed.	· Mostly single-strand breaks (4 min. electrophoresis). · Probably extensive DSB.	· Pregnancy achievement (4 min. electrophoresis time). · Some studies related alkaline Comet to ICSI success using longer times of electrophoresis.
Neutral Comet	Lysis of sperm membranes and extraction of protamines and electrophoresis at neutral pH. DNA breaks migrate towards cathode forming a DNA tail.	· Differentiation of MAR-region specific DSB.	· Technique and analysis are not standardized between laboratories.	· MAR-region specific double strand breaks. · Extensvie DSB.	· First trimester miscarriage risk. · Risk of implantation failure in ICSI cycles. · May be associated to slower embryo kinetics.
Two-tailed Comet	Lysis of sperm membranes and extraction of protamines. First, neutral electrophoresis and, after alkaline denaturation and rotation of slide 90°, alkaline electrophoresis. Sperm present two DNA tails.	· Detection of single and double strand DNA breaks in the same sperm cell.	· Technique not standardized · Difficult interpretation. Requires experienced observer.	· Single-strand breaks. · Extensive DSB. · Not known if MAR-region specific double strand breaks (lack of studies comparing to neutral Comet alone).	· Pregnancy achievement. · Need of human clinical studies regarding ICSI.

HDS: High DNA Stainable sperm; TUNEL: Terminal deoxynucleotidyl transferase dUTP nick end labelling; SCSA: Sperm Chromatin Structure Assay; SCD: Sperm Chromatin Dispersion; ICSI: Intracytoplasmic sperm injection.

On the other hand, Comet assay [41] relies on a DNA decompaction and protein depletion coupled to a single-cell electrophoresis in an agarose micro gel. DNA molecules that contain breaks move towards the cathode and the length of the "comet tail" can be measured to determine the grade of DNA fragmentation at a single cell level. This technique has been applied in multiple different protocols, which usually vary in agarose concentrations and in electrophoresis times [42,43]. As the Comet assay can be performed in alkaline or neutral pH, different types of DNA breaks can be detected (Table 1) (Figure 1): (i) alkaline Comet assay performed in a small electrophoresis time (about four minutes) detect mostly single-strand DNA breaks affecting both toroidal regions and MAR regions in a high number of break points [44,45] and (ii) neutral Comet assay can detect two types of double-strand DNA breaks (Figure 2): (a) extensive DSB, which represent a very small part of total DSB and can be observed as very long comet tails separated from the sperm core; and (b) localized DSB localized and attached to the MAR region, as demonstrated in pulsed-field gel electrophoresis [43–46], being the most common DSB. Although extensive DSB result in longer Comet tails, they cannot be distinguished from localized DSB in a single Comet. However, when a semen sample present high number of sperm cells with extensive DSB (long tails), single-strand DNA damage is also present in a high amount (Ribas-Maynou personal observation). Previous studies had shown that localized DSB represent very few break points in the genome, as long chromatin fibres with a break point in the end can be seen in a detailed neutral Comet image (Figure 2A), which is supported by Kaneko et al., using pulsed field gel electrophoresis [47]. We demonstrated that localized DSB remain attached to the sperm nuclear matrix [45], maybe through a TOP2B or similar protein [45,46], a very important feature taking into account that the nuclear matrix is inherited to the male pronucleus in the zygote [46,48–50], giving a chance to the embryo to repair the DSB.

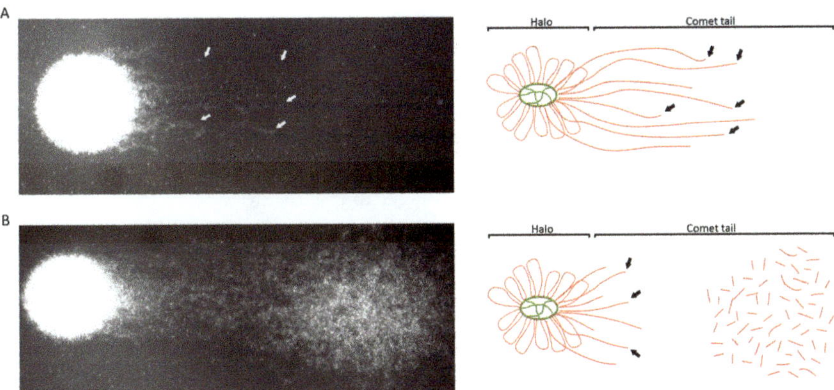

Figure 2. (**A**) Picture and scheme of neutral Comet with localized DSB (double-strand DNA breaks) attached to the nuclear matrix (green). Comet halo consists in non-fragmented chromatin and comet tail is formed by chromatin fibres attached to the nuclear matrix with low number of DNA breaks at the end (arrows). (**B**) Picture and scheme of neutral Comet with extensive DSB. Comet tail is formed by DNA fragments that are not attached to the nuclear matrix. This comet also shows part of localized DNA breaks attached to the MAR region (arrow).

Studies using all the techniques showed that oxidative damage detected by alkaline Comet assay presented a good correlation to TUNEL, SCSA and SCD techniques [23,51,52]. Although these techniques may potentially detect double-strand breaks, a study conducted by our group analysing the same semen samples with five methodologies showed that no correlation was present with the neutral Comet assay [23]. Then, the latter would be the only technique that is able to differentially detect MAR-region double-strand breaks [23,44], whereas TUNEL, SCSA and SCD may detect extensive DSB. A Comet assay variant (two-tailed Comet assay) applying both alkaline and neutral Comet assay in the same slide by turning it 90° between electrophoresis allows to distinguish single and double-strand DNA breaks on the same sperm cell [53]. However, no studies have been performed comparing these techniques and alkaline or neutral Comet assay separately in order to elucidate if double-strand breaks detected in two-tailed Comet assay correspond to MAR region localized DSB.

3. Oxidative DNA Damage, Alkaline Comet Assay and Pregnancy Achievement

Using alkaline Comet assay in different cohorts, an study published in 2012 [43] showed that the extensive single-strand DNA breaks were reversely associated to the achievement of natural pregnancy independently of the neutral Comet results (Figure 1 and Table 1). This was confirmed and compared with TUNEL, SCSA and SCD tests in 2013, demonstrating also that alkaline Comet is the most sensitive technique for the prediction of natural pregnancy achievement [23,43]. Which is also in accordance to the numerous studies from other research groups that find similar association in natural pregnancy using TUNEL, SCSA, SCD and Comet assay tests [5,51,54–58].

Single-strand breaks are produced mainly due to reactive oxygen species (ROS) [42,53,59], which may come from exogenous sources such as environmental toxicants, smoking, alcohol, diet, radiation and so forth or from endogenous sources such as an increase of leukocytes, presence of varicocele or even the ROS generated by mitochondria for the movement of sperm cell [60–62]. Free radicals may cause lipid peroxidation, mitochondrial and nuclear DNA base modifications such as 8-OH-guanine and 8-OH-2′-deoxyguanosine (8-OHdG), an oxidized base adduct that destabilize DNA structure and cause a DNA break [63–65]. This affectation does not find a restriction by DNA condensation and therefore may affect both toroids compacted in protamines and MAR regions compacted in histones [44]. Then, if such an extensive damage happens to the sperm DNA due to oxidative stress, the sperm membranes would also be affected and usually sperm motility is lost. Because of that, a strong negative relation between progressive motility and oxidative damage (single-strand DNA damage) analyzed using TUNEL, SCSA, SCD and alkaline Comet [55,61,66].

As mentioned before in this review, controversial results are found in different studies regarding ICSI outcomes: some of them which found predictive value of oxidative damage [25–28] and other with opposite results [29–33]. If single-strand DNA breaks present a correlation to progressive motility and sperm morphology and ICSI procedures use the most motile sperm cells with better morphology, paternal genome should be free of oxidative damage. In this regard, a work by Gosalvez et al. [67] demonstrated that motile sperm organelle morphology examination (MSOME) selected sperm cells were free of DNA damage analysed by SCD test. Moreover, a work using Comet assay suggested that grade I and II sperm cells present lower incidence of oxidative DNA damage than grade III and IV [68]. These results need to be further confirmed in conventional ICSI sperm selection. However, our data suggest that no relation is present between alkaline Comet and embryo quality, embryo kinetics or implantation [69].

4. Double-Strand DNA Damage, Recurrent Miscarriage and Preimplantation Failure in ICSI Cycles

Analysing the data of the patients and donors with high DSB, a specific profile was observed with low oxidative damage and high neutral comet values in patients with first trimester recurrent miscarriage where all related female factors were discarded and in one subgroup of fertile donors [44]. In a recent study, our group has found that patients with this profile who undergo ICSI treatments produce embryos with a delayed embryo development to blastocyst, which also cause lower implantation rates [69]. Other works also show that double-strand breaks may contribute to a higher implantation failure risk [6,25]. Since implantation failures in ICSI cycles and miscarriages present similar profiles with high DSB, one may think that they might have similar origin. In fact, small number of DNA breaks localized in concrete regions of the genome might induce a cell failure where the affected regions are necessary for the development. In our last study, embryos that achieved implantation presented faster embryo kinetics than those that did not achieve implantation [69]. In fact, faster embryo kinetics had been associated to embryo euploidy [70–72].

DSB are the most lethal alteration that may happen in a zygote, since paternal and maternal pronucleus remain separated in early mammalian embryos and, therefore, no complementary chain would be available for DNA repair [73–75] and a few number of DSB are sufficient to delay cell cycle [76]. It is important to note that paternal double-strand breaks remain attached to the nuclear matrix and probably to other proteins such as TOP2B [20,46,77] and the nuclear matrix is inherited at male pronucleus until first mitotic division [49,78]. This may be crucial at the zygote, because it may give a chance to correctly repair both free ends of the double-strand break. There is a consensus point that oocyte quality may play a role in this DNA repair, since different studies proved that early embryos are able to repair DNA damage [79–84]. In this sense, in patients with DSB, the most significant delay observed in the embryo kinetics was just after fertilization, indicating that DNA repair machinery may be active in this stage [69]. Recent studies in sperm cells demonstrated that MAR regions are required as a scaffold for DNA replication after fertilization [48] and, in somatic cells, nuclear matrix also is involved in transcription, cell regulation and replication [85,86]. In mammals, inducing DSB in sperm cells and used these sperm cells to fertilize eggs observed chromosomal alterations in paternal genome of the embryo and showing also a delay in the first embryo cleavage [17,20,87]. Moreover, studies inducing double-strand DNA breaks in mice sperm through radiation observed a p53 and p21 related response and less number of foetuses [88,89] or less survival of offspring in a dose dependent manner [90].

5. Prevention of DNA Damage

The data presented in the studies referenced before supports that oxidative damage may affect the pregnancy achievement capacity due to misbalanced levels of oxidants/antioxidants [61,91].

The use of antioxidants has been widely applied in subfertile males [92]. Several works demonstrated that they are a positive contribution on sperm count, motility, morphology and also proved that they help reducing oxidative DNA fragmentation [93–96]. Although there are very few studies with randomized and placebo controls, Cochrane review suggests that the use of antioxidants causes from 1.8 to 4.6 fold increase in the chances of achieving a natural pregnancy. However, up to a 6.5 fold increase in miscarriages might be observed [97]. In ICSI treatments, it is still not clear if antioxidants could help on improving pregnancy and birth rates [98–100]. High quality studies including different groups of patients are necessary in order to elucidate the need of antioxidants in ICSI procedures.

Treatments for the reduction of double-strand sperm DNA damage should also reduce the miscarriage risk and the implantation failure risk in ICSI cycles, showing also less delay on embryo kinetics. Until our knowledge, no validated treatment reduce the incidence of MAR-region localized DSB. However, a study conducted in humans in 2006 by Schmid and colleagues demonstrated that men with daily caffeine consumption presented increased values of DSB measured with neutral Comet independently of male age in healthy non-smokers [101]. Caffeine is a known inhibitor of DNA repair, as it has been described that inhibits ATM kinase [102,103] and DNA resection in homologous recombination through Rad51 [104,105]. Also, it has been reported to affect cell cycle at both G1/S and G2/M checkpoints and inducing programmed cell death through p53-dependent pathway [106]. Studies in animals reported that caffeine administration to rats caused an impairment of pregnancy [107]. Other studies inducing DNA strand breaks in sperm cells through radiation and cultivating the oocytes and the produced embryos in caffeine demonstrated that chromosome and chromatid aberrations persist in the zygote, indicating oocyte DNA repair is inhibited by caffeine [17]. Since spermatocytes must produce double-strand breaks through Spo11 in prophase I in order to perform DNA recombination and later, they need to repair these DSB. According to previous results, the consumption of caffeine would impair ATM kinase and/or resection of double-strand breaks [104,105] and may induce that a few double-strand breaks would not be repaired, causing that mature sperm cells present DSB [101]. Further basic studies are needed to explain how a spermatocyte with double-strand breaks can escape the pachytene checkpoint [108,109]. Reducing the incidence of DSB in sperm cell would improve clinical outcomes in terms of miscarriage and implantation in ICSI cycles.

Author Contributions: J.R.-M. and J.B. contributed in manuscript writing and revision.

Funding: This research was funded by the European Regional Development Fund and Instituto de Salud Carlos III (Economy, Industry and Competitiveness Ministry, Madrid, Spain; Project PI14/Q1 00119) and Generalitat de Catalunya (Project 2017SGR1796).

Conflicts of Interest: The authors declare no conflict of interest.

References

1. Thoma, M.E.; McLain, A.C.; Louis, J.F.; King, R.B.; Trumble, A.C.; Sundaram, R.; Buck Louis, G.M. Prevalence of infertility in the United States as estimated by the current duration approach and a traditional constructed approach. *Fertil. Steril.* **2013**, *99*, 1324–1331. [CrossRef] [PubMed]
2. Louis, J.F.; Thoma, M.E.; Sørensen, D.N.; McLain, A.C.; King, R.B.; Sundaram, R.; Keiding, N.; Buck Louis, G.M. The prevalence of couple infertility in the United States from a male perspective: Evidence from a nationally representative sample. *Andrology* **2013**, *1*, 741–748. [CrossRef] [PubMed]
3. Baird, D.T.; Collins, J.; Egozcue, J.; Evers, L.H.; Gianaroli, L.; Leridon, H.; Sunde, A.; Templeton, A.; Van Steirteghem, A.; Cohen, J.; et al. Fertility and ageing. *Hum. Reprod. Update* **2005**, *11*, 261–276. [PubMed]
4. De Kretser, D.M. Male infertility. *Lancet* **1997**, *349*, 787–790. [CrossRef]
5. Lewis, S.E.M.; Simon, L. Clinical implications of sperm DNA damage. *Hum. Fertil. (Camb.)* **2010**, *13*, 201–207. [CrossRef] [PubMed]
6. Simon, L.; Murphy, K.; Shamsi, M.B.; Liu, L.; Emery, B.; Aston, K.I.; Hotaling, J.; Carrell, D.T. Paternal influence of sperm DNA integrity on early embryonic development. *Hum. Reprod.* **2014**, *29*, 2402–2412. [CrossRef] [PubMed]
7. Cooper, T.G.; Noonan, E.; von Eckardstein, S.; Auger, J.; Baker, H.W.G.; Behre, H.M.; Haugen, T.B.; Kruger, T.; Wang, C.; Mbizvo, M.T.; et al. World Health Organization reference values for human semen characteristics. *Hum. Reprod. Update* **2010**, *16*, 231–245. [CrossRef] [PubMed]
8. Yu, S.; Rubin, M.; Geevarughese, S.; Pino, J.S.; Rodriguez, H.F.; Asghar, W. Emerging technologies for home-based semen analysis. *Andrology* **2018**, *6*, 10–19. [CrossRef] [PubMed]
9. Wang, C.; Swerdloff, R.S. Limitations of semen analysis as a test of male fertility and anticipated needs from newer tests. *Fertil. Steril.* **2014**, *102*, 1502–1507. [CrossRef] [PubMed]
10. Kupka, M.S.; D'Hooghe, T.; Ferraretti, A.P.; De Mouzon, J.; Erb, K.; Castilla, J.A.; Calhaz-Jorge, C.; De Geyter, C.; Goossens, V. Assisted reproductive technology in Europe, 2011: Results generated from European registers by ESHRE. *Hum. Reprod.* **2015**, *31*, 233–248.
11. Love, C.C.; Kenney, R.M. Scrotal heat stress induces altered sperm chromatin structure associated with a decrease in protamine disulfide bonding in the stallion. *Biol. Reprod.* **1999**, *60*, 615–620. [CrossRef] [PubMed]
12. Keeney, S.; Lange, J.; Mohibullah, N. Self-organization of meiotic recombination initiation: General principles and molecular pathways. *Annu. Rev. Genet.* **2014**, *48*, 187–214. [CrossRef] [PubMed]
13. Cooper, T.J.; Wardell, K.; Garcia, V.; Neale, M.J. Homeostatic regulation of meiotic DSB formation by ATM/ATR. *Exp. Cell Res.* **2014**, *329*, 124–131. [CrossRef] [PubMed]
14. Alberts, B.; Johnson, A.; Lewis, J.; Raff, M.; Roberts, K.; Walter, P. *Molecular Biology of the Cell*, 4th ed.; Garland Sc.: New York, NY, USA, 2002.
15. Ward, W.S.; Coffey, D.S. Specific organization of genes in relation to the sperm nuclear matrix. *Biochem. Biophys. Res. Commun.* **1990**, *173*, 20–25. [CrossRef]
16. Ward, W.S. Function of sperm chromatin structural elements in fertilization and development. *Mol. Hum. Reprod.* **2010**, *16*, 30–36. [CrossRef] [PubMed]
17. Genescà, A.; Caballín, M.R.; Miró, R.; Benet, J.; Germà, J.R.; Egozcue, J. Repair of human sperm chromosome aberrations in the hamster egg. *Hum. Genet.* **1992**, *89*, 181–186. [CrossRef]
18. Oh, J.; Symington, L. Role of the Mre11 complex in preserving genome integrity. *Genes (Basel)* **2018**, *9*, 589. [CrossRef]

19. Barnes, J.L.; Zubair, M.; John, K.; Poirier, M.C.; Martin, F.L. Carcinogens and DNA damage. *Biochem. Soc. Trans.* **2018**, *46*, 1213–1224. [CrossRef]

20. Gawecka, J.E.; Marh, J.; Ortega, M.; Yamauchi, Y.; Ward, M.A.; Ward, W.S. Mouse zygotes respond to severe sperm DNA damage by delaying paternal DNA replication and embryonic development. *PLoS ONE* **2013**, *8*, e56385. [CrossRef]

21. Lewis, S.E.M.; John Aitken, R.; Conner, S.J.; De Iuliis, G.; Evenson, D.P.; Henkel, R.; Giwercman, A.; Gharagozloo, P. The impact of sperm DNA damage in assisted conception and beyond: Recent advances in diagnosis and treatment. *Reprod. Biomed. Online* **2013**, *27*, 325–337. [CrossRef]

22. Evenson, D.P. Sperm chromatin structure assay (SCSA®). *Methods Mol. Biol.* **2013**, *927*, 147–164. [PubMed]

23. Ribas-Maynou, J.; García-Peiró, A.; Fernández-Encinas, A.; Abad, C.; Amengual, M.J.; Prada, E.; Navarro, J.; Benet, J. Comprehensive analysis of sperm DNA fragmentation by five different assays: TUNEL assay, SCSA, SCD test and alkaline and neutral Comet assay. *Andrology* **2013**, *1*, 715–722. [CrossRef]

24. Peluso, G.; Palmieri, A.; Cozza, P.P.; Morrone, G.; Verze, P.; Longo, N.; Mirone, V. The study of spermatic DNA fragmentation and sperm motility in infertile subjects. *Arch. Ital. di Urol. e Androl.* **2013**, *85*, 8. [CrossRef] [PubMed]

25. Garolla, A.; Cosci, I.; Bertoldo, A.; Sartini, B.; Boudjema, E.; Foresta, C. DNA double strand breaks in human spermatozoa can be predictive for assisted reproductive outcome. *Reprod. Biomed. Online* **2015**, *31*, 100–107. [CrossRef] [PubMed]

26. Simon, L.; Brunborg, G.; Stevenson, M.; Lutton, D.; McManus, J.; Lewis, S.E.M. Clinical significance of sperm DNA damage in assisted reproduction outcome. *Hum. Reprod.* **2010**, *25*, 1594–1608. [CrossRef] [PubMed]

27. Simon, L.; Proutski, I.; Stevenson, M.; Jennings, D.; McManus, J.; Lutton, D.; Lewis, S.E.M. Sperm DNA damage has a negative association with live-birth rates after IVF. *Reprod. Biomed. Online* **2013**, *26*, 68–78. [CrossRef] [PubMed]

28. Meseguer, M.; Santiso, R.; Garrido, N.; García-Herrero, S.; Remohí, J.; Fernandez, J.L. Effect of sperm DNA fragmentation on pregnancy outcome depends on oocyte quality. *Fertil. Steril.* **2011**, *95*, 124–128. [CrossRef]

29. Esbert, M.; Pacheco, A.; Vidal, F.; Florensa, M.; Riqueros, M.; Ballesteros, A.; Garrido, N.; Calderón, G. Impact of sperm DNA fragmentation on the outcome of IVF with own or donated oocytes. *Reprod. Biomed. Online* **2011**, *23*, 704–710. [CrossRef]

30. Thomson, L.K.; Zieschang, J.-A.; Clark, A.M. Oxidative deoxyribonucleic acid damage in sperm has a negative impact on clinical pregnancy rate in intrauterine insemination but not intracytoplasmic sperm injection cycles. *Fertil. Steril.* **2011**, *96*, 843–847. [CrossRef]

31. Pregl Breznik, B.; Kovačič, B.; Vlaisavljević, V. Are sperm DNA fragmentation, hyperactivation and hyaluronan-binding ability predictive for fertilization and embryo development in in vitro fertilization and intracytoplasmic sperm injection? *Fertil. Steril.* **2013**, *99*, 1233–1241. [CrossRef]

32. Anifandis, G.; Bounartzi, T.; Messini, C.I.; Dafopoulos, K.; Markandona, R.; Sotiriou, S.; Tzavella, A.; Messinis, I.E. Sperm DNA fragmentation measured by Halosperm does not impact on embryo quality and ongoing pregnancy rates in IVF/ICSI treatments. *Andrologia* **2015**, *47*, 295–302. [CrossRef] [PubMed]

33. Haghpanah, T.; Salehi, M.; Ghaffari Novin, M.; Masteri Farahani, R.; Fadaei-Fathabadi, F.; Dehghani-Mohammadabadi, M.; Azimi, H. Does sperm DNA fragmentation affect the developmental potential and the incidence of apoptosis following blastomere biopsy? *Syst. Biol. Reprod. Med.* **2016**, *62*, 1–10. [CrossRef] [PubMed]

34. ASRM. The clinical utility of sperm DNA integrity testing: A guideline. *Fertil. Steril.* **2013**, *99*, 673–677. [CrossRef] [PubMed]

35. Gorczyca, W.; Gong, J.; Darzynkiewicz, Z. Detection of DNA strand breaks in individual apoptotic cells by the in situ terminal deoxynucleotidyl transferase and nick translation assays. *Cancer Res.* **1993**, *53*, 1945–1951. [PubMed]

36. Sharma, R.; Masaki, J.; Agarwal, A. Sperm DNA fragmentation analysis using the TUNEL assay. *Methods Mol. Biol.* **2013**, *927*, 121–136. [PubMed]

37. Mitchell, L.A.; De Iuliis, G.N.; Aitken, R.J. The TUNEL assay consistently underestimates DNA damage in human spermatozoa and is influenced by DNA compaction and cell vitality: Development of an improved methodology. *Int. J. Androl.* **2011**, *34*, 2–13. [CrossRef]

38. Domínguez-Fandos, D.; Camejo, M.I.; Ballescà, J.L.; Oliva, R. Human sperm DNA fragmentation: correlation of TUNEL results as assessed by flow cytometry and optical microscopy. *Cytometry. A* **2007**, *71*, 1011–1018. [CrossRef]

39. Evenson, D.; Jost, L. Sperm chromatin structure assay is useful for fertility assessment. *Methods Cell Sci.* **2000**, *22*, 169–189. [CrossRef]

40. Fernández, J.L.; Muriel, L.; Goyanes, V.; Segrelles, E.; Gosálvez, J.; Enciso, M.; LaFromboise, M.; De Jonge, C. Simple determination of human sperm DNA fragmentation with an improved sperm chromatin dispersion test. *Fertil. Steril.* **2005**, *84*, 833–842. [CrossRef]

41. Singh, N.P.; McCoy, M.T.; Tice, R.R.; Schneider, E.L. A simple technique for quantitation of low levels of DNA damage in individual cells. *Exp. Cell Res.* **1988**, *175*, 184–191. [CrossRef]

42. Simon, L.; Carrell, D.T. Sperm DNA damage measured by comet assay. *Methods Mol. Biol.* **2013**, *927*, 137–146. [CrossRef] [PubMed]

43. Ribas-Maynou, J.; Garca-Peiro, A.; Abad, C.; Amengual, M.J.; Navarro, J.; Benet, J. Alkaline and neutral Comet assay profiles of sperm DNA damage in clinical groups. *Hum. Reprod.* **2012**, *27*, 652–658. [CrossRef]

44. Ribas-Maynou, J.; García-Peiró, A.; Fernandez-Encinas, A.; Amengual, M.J.; Prada, E.; Cortés, P.; Navarro, J.; Benet, J. Double stranded sperm DNA breaks, measured by comet assay, are associated with unexplained recurrent miscarriage in couples without a female factor. *PLoS ONE* **2012**, *7*. [CrossRef]

45. Ribas-Maynou, J.; Gawecka, J.E.; Benet, J.; Ward, W.S. Double-stranded DNA breaks hidden in the neutral Comet assay suggest a role of the sperm nuclear matrix in DNA integrity maintenance. *Mol. Hum. Reprod.* **2014**, *20*, 330–340. [CrossRef] [PubMed]

46. Gawecka, J.E.; Ribas-Maynou, J.; Benet, J.; Ward, W.S. A model for the control of DNA integrity by the sperm nuclear matrix. *Asian J. Androl.* **2015**, *17*.

47. Kaneko, S.; Yoshida, J.; Ishikawa, H.; Takamatsu, K. Single-cell pulsed-field gel electrophoresis to detect the early stage of DNA fragmentation in human sperm nuclei. *PLoS ONE* **2012**, *7*, e42257. [CrossRef] [PubMed]

48. Shaman, J.A.; Yamauchi, Y.; Ward, W.S. Function of the sperm nuclear matrix. *Arch. Androl.* **2007**, *53*, 135–140. [CrossRef]

49. Mohar, I.; Szczygiel, M.A.; Yanagimachi, R.; Ward, W.S. Sperm nuclear halos can transform into normal chromosomes after injection into oocytes. *Mol. Reprod. Dev.* **2002**, *62*, 416–420. [CrossRef] [PubMed]

50. Shaman, J.A.; Yamauchi, Y.; Ward, W.S. The sperm nuclear matrix is required for paternal DNA replication. *J. Cell. Biochem.* **2007**, *102*, 680–688. [CrossRef]

51. Simon, L.; Liu, L.; Murphy, K.; Ge, S.; Hotaling, J.; Aston, K.I.; Emery, B.; Carrell, D.T. Comparative analysis of three sperm DNA damage assays and sperm nuclear protein content in couples undergoing assisted reproduction treatment. *Hum. Reprod.* **2014**, *29*, 904–917. [CrossRef]

52. Chohan, K.R.; Griffin, J.T.; Lafromboise, M.; De Jonge, C.J.; Carrell, D.T. Comparison of chromatin assays for DNA fragmentation evaluation in human sperm. *J. Androl.* **2006**, *27*, 53–59. [CrossRef] [PubMed]

53. Enciso, M.; Sarasa, J.; Agarwal, A.; Fernández, J.L.; Gosálvez, J. A two-tailed Comet assay for assessing DNA damage in spermatozoa. *Reprod. Biomed. Online* **2009**, *18*, 609–616. [CrossRef]

54. Saleh, R.A.; Agarwal, A.; Nelson, D.R.; Nada, E.A.; El-Tonsy, M.H.; Alvarez, J.G.; Thomas, A.J.; Sharma, R.K. Increased sperm nuclear DNA damage in normozoospermic infertile men: A prospective study. *Fertil. Steril.* **2002**, *78*, 313–318. [CrossRef]

55. Belloc, S.; Benkhalifa, M.; Cohen-Bacrie, M.; Dalleac, A.; Amar, E.; Zini, A. Sperm deoxyribonucleic acid damage in normozoospermic men is related to age and sperm progressive motility. *Fertil. Steril.* **2014**, *101*, 1588–1593. [CrossRef] [PubMed]

56. Cho, C.-L.; Agarwal, A. Role of sperm DNA fragmentation in male factor infertility: A systematic review. *Arab J. Urol.* **2018**, *16*, 21–34. [CrossRef]

57. Santi, D.; Spaggiari, G.; Simoni, M. Sperm DNA fragmentation index as a promising predictive tool for male infertility diagnosis and treatment management – meta-analyses. *Reprod. Biomed. Online* **2018**, *37*, 315–326. [CrossRef] [PubMed]

58. Simon, L.; Lutton, D.; McManus, J.; Lewis, S.E.M. Sperm DNA damage measured by the alkaline Comet assay as an independent predictor of male infertility and in vitro fertilization success. *Fertil. Steril.* **2011**, *95*, 652–657. [CrossRef]

59. Agarwal, A.; Prabakaran, S.A. Mechanism, measurement and prevention of oxidative stress in male reproductive physiology. *Indian J. Exp. Biol.* **2005**, *43*, 963–974.

60. Agarwal, A.; Virk, G.; Ong, C.; du Plessis, S.S. Effect of oxidative stress on male reproduction. *World J. Mens. Health* **2014**, *32*, 1. [CrossRef]

61. Aitken, R.J.; De Iuliis, G.N. On the possible origins of DNA damage in human spermatozoa. *Mol. Hum. Reprod.* **2010**, *16*, 3–13. [CrossRef]

62. Sakkas, D.; Alvarez, J.G. Sperm DNA fragmentation: mechanisms of origin, impact on reproductive outcome and analysis. *Fertil. Steril.* **2010**, *93*, 1027–1036. [CrossRef] [PubMed]

63. Lord, T.; Aitken, R.J. Fertilization stimulates 8-hydroxy-2′-deoxyguanosine repair and antioxidant activity to prevent mutagenesis in the embryo. *Dev. Biol.* **2015**, *406*, 1–13. [CrossRef] [PubMed]

64. De Iuliis, G.N.; Thomson, L.K.; Mitchell, L.A.; Finnie, J.M.; Koppers, A.J.; Hedges, A.; Nixon, B.; Aitken, R.J. DNA damage in human spermatozoa is highly correlated with the efficiency of chromatin remodeling and the formation of 8-hydroxy-2′-deoxyguanosine, a marker of oxidative stress. *Biol. Reprod.* **2009**, *81*, 517–524. [CrossRef] [PubMed]

65. Santiso, R.; Tamayo, M.; Gosálvez, J.; Meseguer, M.; Garrido, N.; Fernández, J.L. Simultaneous determination in situ of DNA fragmentation and 8-oxoguanine in human sperm. *Fertil. Steril.* **2010**, *93*, 314–318. [CrossRef] [PubMed]

66. Evgeni, E.; Lymberopoulos, G.; Touloupidis, S.; Asimakopoulos, B. Sperm nuclear DNA fragmentation and its association with semen quality in Greek men. *Andrologia* **2015**, *47*, 1166–1174. [CrossRef] [PubMed]

67. Gosálvez, J.; Migueles, B.; López-fernández, C.; Sanchéz-martín, F.; Sáchez-martín, P. Single sperm selection and DNA fragmentation analysis: The case of MSOME/IMSI. *Nat. Sci. Res.* **2013**, *5*, 7–14. [CrossRef]

68. Pastuszek, E.; Kiewisz, J.; Skowronska, P.; Liss, J.; Lukaszuk, M.; Bruszczynska, A.; Jakiel, G.; Lukaszuk, K. An investigation of the potential effect of sperm nuclear vacuoles in human spermatozoa on DNA fragmentation using a neutral and alkaline Comet assay. *Andrology* **2017**, *5*, 392–398. [CrossRef]

69. Casanovas, A.; Ribas-Maynou, J.; Lara-Cerrillo, S.; Jimenez-Macedo, A.R.; Hortal, O.; Benet, J.; Carrera, J.; Garcia-Peiró, A. Double-stranded sperm DNA damage is a cause of delay in embryo development and can impair implantation rates. *Fertil. Steril.* **2018**, in press.

70. Campbell, A.; Fishel, S.; Laegdsmand, M. Aneuploidy is a key causal factor of delays in blastulation: author response to "A cautionary note against aneuploidy risk assessment using time-lapse imaging". *Reprod. Biomed. Online* **2014**, *28*, 279–283. [CrossRef]

71. Basile, N.; del Carmen Nogales, M.; Bronet, F.; Florensa, M.; Riqueiros, M.; Rodrigo, L.; García-Velasco, J.; Meseguer, M. Increasing the probability of selecting chromosomally normal embryos by time-lapse morphokinetics analysis. *Fertil. Steril.* **2014**, *101*, 699–704. [CrossRef]

72. Campbell, A.; Fishel, S.; Bowman, N.; Duffy, S.; Sedler, M.; Thornton, S. Retrospective analysis of outcomes after IVF using an aneuploidy risk model derived from time-lapse imaging without PGS. *Reprod. Biomed. Online* **2013**, *27*, 140–146. [CrossRef] [PubMed]

73. Bernstein, K.A.; Rothstein, R. At loose ends: Resecting a double-strand break. *Cell* **2009**, *137*, 807–810. [CrossRef] [PubMed]

74. Price, B.D.; D'Andrea, A.D. Chromatin remodeling at DNA double-strand breaks. *Cell* **2013**, *152*, 1344–1354. [CrossRef] [PubMed]

75. Reichmann, J.; Nijmeijer, B.; Hossain, M.J.; Eguren, M.; Schneider, I.; Politi, A.Z.; Roberti, M.J.; Hufnagel, L.; Hiiragi, T.; Ellenberg, J. Dual-spindle formation in zygotes keeps parental genomes apart in early mammalian embryos. *Science* **2018**, *361*, 189–193. [CrossRef] [PubMed]

76. Van den Berg, J.; G Manjón, A.; Kielbassa, K.; Feringa, F.M.; Freire, R.; Medema, R.H. A limited number of double-strand DNA breaks is sufficient to delay cell cycle progression. *Nucleic Acids Res.* **2018**, *46*, 10132–10144. [CrossRef] [PubMed]

77. Yamauchi, Y.; Shaman, J.A.; Ward, W.S. Topoisomerase II-mediated breaks in spermatozoa cause the specific degradation of paternal DNA in fertilized oocytes. *Biol. Reprod.* **2007**, *76*, 666–672. [CrossRef]

78. Ward, W.S.; Kimura, Y.; Yanagimachi, R. An intact sperm nuclear matrix may be necessary for the mouse paternal genome to participate in embryonic development. *Biol. Reprod.* **1999**, *60*, 702–706. [CrossRef]

79. Zheng, P.; Schramm, R.D.; Latham, K.E. Developmental regulation and in vitro culture effects on expression of DNA repair and cell cycle checkpoint control genes in rhesus monkey oocytes and embryos. *Biol. Reprod.* **2005**, *72*, 1359–1369. [CrossRef] [PubMed]

80. Marchetti, F.; Bishop, J.; Gingerich, J.; Wyrobek, A.J. Meiotic interstrand DNA damage escapes paternal repair and causes chromosomal aberrations in the zygote by maternal misrepair. *Sci. Rep.* **2015**, *5*, 7689. [CrossRef] [PubMed]

81. Menezo, Y.; Russo, G.; Tosti, E.; El Mouatassim, S.; Benkhalifa, M. Expression profile of genes coding for DNA repair in human oocytes using pangenomic microarrays, with a special focus on ROS linked decays. *J. Assist. Reprod. Genet.* **2007**, *24*, 513–520. [CrossRef]

82. Jaroudi, S.; Kakourou, G.; Cawood, S.; Doshi, A.; Ranieri, D.M.; Serhal, P.; Harper, J.C.; SenGupta, S.B. Expression profiling of DNA repair genes in human oocytes and blastocysts using microarrays. *Hum. Reprod.* **2009**, *24*, 2649–2655. [CrossRef] [PubMed]

83. Martin, J.H.; Bromfield, E.G.; Aitken, R.J.; Lord, T.; Nixon, B. Double strand break DNA repair occurs via non-homologous end-joining in mouse MII oocytes. *Sci. Rep.* **2018**, *8*, 9685. [CrossRef] [PubMed]

84. García-Rodríguez, A.; Gosálvez, J.; Agarwal, A.; Roy, R.; Johnston, S. DNA Damage and repair in human reproductive cells. *Int. J. Mol. Sci.* **2018**, *20*, 31. [CrossRef] [PubMed]

85. Pardoll, D.M.; Vogelstein, B.; Coffey, D.S. A fixed site of DNA replication in eucaryotic cells. *Cell* **1980**, *19*, 527–536. [CrossRef]

86. Wilson, R.H.C.; Coverley, D. Relationship between DNA replication and the nuclear matrix. *Genes Cells* **2013**, *18*, 17–31. [CrossRef] [PubMed]

87. Tusell, L.; Latre, L.; Ponsa, I.; Miró, R.; Egozcue, J.; Genescà, A. Capping of radiation-induced DNA breaks in mouse early embryos. *J. Radiat. Res.* **2004**, *45*, 415–422. [CrossRef] [PubMed]

88. Adiga, S.K.; Toyoshima, M.; Shiraishi, K.; Shimura, T.; Takeda, J.; Taga, M.; Nagai, H.; Kumar, P.; Niwa, O. p21 provides stage specific DNA damage control to preimplantation embryos. *Oncogene* **2007**, *26*, 6141–6149. [CrossRef]

89. Toyoshima, M. Analysis of p53 dependent damage response in sperm-irradiated mouse embryos. *J. Radiat. Res.* **2009**, *50*, 11–17. [CrossRef]

90. Kumar, D.; Upadhya, D.; Salian, S.R.; Rao, S.B.S.; Kalthur, G.; Kumar, P.; Adiga, S.K. The extent of paternal sperm DNA damage influences early post-natal survival of first generation mouse offspring. *Eur. J. Obstet. Gynecol. Reprod. Biol.* **2013**, *166*, 164–167. [CrossRef]

91. Majzoub, A.; Agarwal, A. Antioxidant therapy in idiopathic oligoasthenoteratozoospermia. *Indian J. Urol.* **2017**, *33*, 207.

92. Gharagozloo, P.; Aitken, R.J. The role of sperm oxidative stress in male infertility and the significance of oral antioxidant therapy. *Hum. Reprod.* **2011**, *26*, 1628–1640. [CrossRef] [PubMed]

93. Menezo, Y.; Evenson, D.; Cohen, M.; Dale, B. Effect of Antioxidants on Sperm Genetic Damage. In *Genetic Damage in Human Spermatozoa*; Advances in Experimental Medicine and Biology; Springer: New York, NY, USA, 2014; Volume 791, pp. 173–189.

94. Zini, A.; San Gabriel, M.; Baazeem, A. Antioxidants and sperm DNA damage: A clinical perspective. *J. Assist. Reprod. Genet.* **2009**, *26*, 427–432. [CrossRef] [PubMed]

95. Rahman, S.; Huang, Y.; Zhu, L.; Feng, S.; Khan, I.; Wu, J.; Li, Y.; Wang, X. therapeutic role of green tea polyphenols in improving fertility: A review. *Nutrients* **2018**, *10*, 834. [CrossRef] [PubMed]

96. Imamovic Kumalic, S.; Pinter, B. Review of clinical trials on effects of oral antioxidants on basic semen and other parameters in idiopathic oligoasthenoteratozoospermia. *Biomed Res. Int.* **2014**, *2014*, 1–11. [CrossRef] [PubMed]

97. Showell, M.G.; Mackenzie-Proctor, R.; Brown, J.; Yazdani, A.; Stankiewicz, M.T.; Hart, R.J. Antioxidants for male subfertility. *Cochrane Database Syst. Rev.* **2014**, CD007411. [CrossRef] [PubMed]

98. Agarwal, A.; Durairajanayagam, D.; du Plessis, S.S. Utility of antioxidants during assisted reproductive techniques: An evidence based review. *Reprod. Biol. Endocrinol.* **2014**, *12*, 112. [CrossRef]

99. Greco, E.; Romano, S.; Iacobelli, M.; Ferrero, S.; Baroni, E.; Minasi, M.G.; Ubaldi, F.; Rienzi, L.; Tesarik, J. ICSI in cases of sperm DNA damage: Beneficial effect of oral antioxidant treatment. *Hum. Reprod.* **2005**, *20*, 2590–2594. [CrossRef] [PubMed]

100. Tremellen, K.; Miari, G.; Froiland, D.; Thompson, J. A randomised control trial examining the effect of an antioxidant (Menevit) on pregnancy outcome during IVF-ICSI treatment. *Aust. N. Z. J. Obstet. Gynaecol.* **2007**, *47*, 216–221. [CrossRef] [PubMed]

101. Schmid, T.E.; Eskenazi, B.; Baumgartner, A.; Marchetti, F.; Young, S.; Weldon, R.; Anderson, D.; Wyrobek, A.J. The effects of male age on sperm DNA damage in healthy non-smokers. *Hum. Reprod.* **2006**, *22*, 180–187, Epub 19 October 2006. [CrossRef]
102. Meng, A.; Jiang, L. Induction of G2/M arrest by pseudolaric acid B is mediated by activation of the ATM signaling pathway. *Acta Pharmacol. Sin.* **2009**, *30*, 442–450. [CrossRef]
103. Sabisz, M.; Skladanowski, A. Modulation of cellular response to anticancer treatment by caffeine: inhibition of cell cycle checkpoints, DNA repair and more. *Curr. Pharm. Biotechnol.* **2008**, *9*, 325–336. [CrossRef] [PubMed]
104. Tsabar, M.; Eapen, V.V.; Mason, J.M.; Memisoglu, G.; Waterman, D.P.; Long, M.J.; Bishop, D.K.; Haber, J.E. Caffeine impairs resection during DNA break repair by reducing the levels of nucleases Sae2 and Dna2. *Nucleic Acids Res.* **2015**, *43*, 6889–6901. [CrossRef] [PubMed]
105. Tsabar, M.; Mason, J.M.; Chan, Y.-L.; Bishop, D.K.; Haber, J.E. Caffeine inhibits gene conversion by displacing Rad51 from ssDNA. *Nucleic Acids Res.* **2015**, *43*, 6902–6918. [CrossRef] [PubMed]
106. Bode, A.M.; Dong, Z. The enigmatic effects of caffeine in cell cycle and cancer. *Cancer Lett.* **2007**, *247*, 26–39. [CrossRef] [PubMed]
107. Moskovtsev, S.I.; Willis, J.; White, J.; Mullen, J.B.M. Disruption of telomere-telomere interactions associated with DNA damage in human spermatozoa. *Syst. Biol. Reprod. Med.* **2010**, *56*, 407–412. [CrossRef] [PubMed]
108. Roeder, G.S.; Bailis, J.M. The pachytene checkpoint. *Trends Genet.* **2000**, *16*, 395–403. [CrossRef]
109. Tsubouchi, H.; Argunhan, B.; Tsubouchi, T. Exiting prophase I: No clear boundary. *Curr. Genet.* **2018**, *64*, 423–427. [CrossRef]

Article

Automated Nuclear Cartography Reveals Conserved Sperm Chromosome Territory Localization across 2 Million Years of Mouse Evolution

Benjamin Matthew Skinner [1,*], Joanne Bacon [1], Claudia Cattoni Rathje [2], Erica Lee Larson [3,4], Emily Emiko Konishi Kopania [4], Jeffrey Martin Good [4], Nabeel Ahmed Affara [1] and Peter James Ivor Ellis [2,*]

[1] Department of Pathology, University of Cambridge, Cambridge CB2 1QP, UK; jb552@cam.ac.uk (J.B.); na106@cam.ac.uk (N.A.A.)
[2] School of Biosciences, University of Kent, Canterbury CT2 7NJ, UK; C.C.Rathje@kent.ac.uk
[3] Department of Biological Sciences, University of Denver, Denver, CO 80208, USA; erica.larson@du.edu
[4] Division of Biological Sciences, University of Montana, MT 59812, USA; emily.kopania@umconnect.umt.edu (E.E.K.K.); jeffrey.good@mso.umt.edu (J.M.G.)
[*] Correspondence: bms41@cam.ac.uk (B.M.S.); P.J.I.Ellis@kent.ac.uk (P.J.I.E.)

Received: 30 December 2018; Accepted: 28 January 2019; Published: 1 February 2019

Abstract: Measurements of nuclear organization in asymmetric nuclei in 2D images have traditionally been manual. This is exemplified by attempts to measure chromosome position in sperm samples, typically by dividing the nucleus into zones, and manually scoring which zone a fluorescence in-situ hybridisation (FISH) signal lies in. This is time consuming, limiting the number of nuclei that can be analyzed, and prone to subjectivity. We have developed a new approach for automated mapping of FISH signals in asymmetric nuclei, integrated into an existing image analysis tool for nuclear morphology. Automatic landmark detection defines equivalent structural regions in each nucleus, then dynamic warping of the FISH images to a common shape allows us to generate a composite of the signal within the entire cell population. Using this approach, we mapped the positions of the sex chromosomes and two autosomes in three mouse lineages (*Mus musculus domesticus*, *Mus musculus musculus* and *Mus spretus*). We found that in all three, chromosomes 11 and 19 tend to interact with each other, but are shielded from interactions with the sex chromosomes. This organization is conserved across 2 million years of mouse evolution.

Keywords: nuclear organization; sperm; morphometrics; chromosome painting

---

1. Introduction

Studies of the sub-nuclear localisation of chromatin often use fluorescence in-situ hybridisation (FISH) to detect DNA or RNA, or immunostaining to detect proteins. The images are subsequently analysed either manually or using some automated analysis tool. If the nucleus is circular or elliptical, it is commonly divided into concentric shells of equal area and the proportion of signal in each shell is measured (e.g., [1–3]). This has been amenable to automation, allowing analysis of thousands of cells, which, with appropriate statistical treatment, can yield valuable data at a scale that is still beyond the scope of 3D imaging techniques in time and cost.

However, if the nucleus is asymmetric, such as in sperm, a shell analysis is not sufficient. Frequently, nuclei are manually divided into geometric regions, and the number of nuclei with signals in each region are counted. For example, in spatulate sperm, such as pig or human, positions of loci are located into anterior, medial and posterior regions [4–6], or measured by proportional position along each axis [7]. Rodent sperm have a more interesting, falciform, hooked shape: They have two axes of asymmetry, the anterior-posterior and the dorsal-ventral axis. This means that the location of a FISH signal can—in principle—be unambiguously localised and compared between nuclei. The determination of chromosome position is still manual, with more regions of the nucleus into which a signal may be assigned [8,9], or described without quantitation [10]. This is both time-consuming, and subjective, limiting the numbers of nuclei that can be analysed.

The positions of chromosomes or other loci in gametes (particularly sperm) is of great interest due to both the association of nuclear organisation with fertility in the clinic, in agriculture, and in evolutionary biology. Chromosome position has been linked with infertility in human males; men presenting with fertility problems have less consistent chromosome territories than healthy men [11–13]. Similarly, in farm animals, studies of nuclear organisation have discovered conserved sperm chromosome territories in boars [4], and wider evolutionary studies have shown conservation of some chromosomes, such as the X, from eutherian mammals to marsupial mammals and monotremes [14].

Newer sequencing-based approaches, such as Hi-C are being used to produce 3D maps of chromatin structure across multiple and even single nuclei [15–17]. Validating these results by microscopy is harder due to the number of cells that must be analysed, yet is necessary for our understanding of how chromatin patterns seen across millions of cells relate to chromatin structure within an individual nucleus. Three-dimensional imaging such as confocal microscopy provides high quality position information, but is time-consuming and costly in comparison to 2D fluorescence imaging.

Given this, there is a need to quickly and robustly assay nuclear organisation in 2D fluorescence microscopy images with greater precision than is currently available. Here, we demonstrate the use of automatic landmark detection in nuclei to rapidly localise, aggregate and compare nuclear signals without need for precise detection of the signal boundaries, or extensive manual thresholding and curation. We use this method to investigate the conservation of nuclear organisation between three mouse lineages, *Mus musculus musculus*, *Mus musculus domesticus* and *Mus spretus*. Of these, *M. spretus* has a notably different nuclear shape [18] to the others, being shorter and wider, allowing us to test whether chromosome position is conserved across structurally equivalent regions.

2. Materials and Methods

2.1. Sample Collection

We collected sperm from wild-derived inbred mouse strains *Mus musculus musculus* (PWK/PhJ), *M. m. domesticus* (LEWES/EiJ) and *Mus spretus* (STF). All animal procedures were subject to local ethical review by the University of Montana Institute for Animal Care and Use Committee (protocol identification number 002-13JGDBS-011613, approved January 16, 2013). Animals were bred at the University of Montana from mice purchased from Jackson Laboratories (Bar Harbor, ME, USA) or were acquired from Francois Bonhomme (University of Montpellier, France). Animals were housed singly or in small groups, sacrificed via CO_2 followed by cervical dislocation, and tissues were collected post mortem for analysis. Sperm were collected and fixed in 3:1 methanol-acetic acid as previously described [18].

2.2. Fluorescence In-Situ Hybridisation (FISH)

Fixed sperm were dropped on poly-lysine slides, air-dried, and aged at 70 °C for one hour. Sperm were swelled in 10 mM DTT in 0.1 M Tris-Hcl for 30 min at room temperature (RT). Slides were rinsed in 2 × saline sodium citrate (SSC) and dehydrated through an ethanol series (70%, 80%, 100%, 2 min at RT). Chromatin was relaxed by incubating slides in 0.1 mg/mL pepsin in 0.01 N HCl at 37 °C for 20 min. Nuclei were permeabilized in 0.5% IGEPAL CA-630, 0.5% Triton-X-100 at 4 °C for 30 min, and dehydrated through an ethanol series. Slides and chromosome paints for chrX, Y, 11 and 19 (Cytocell, Cambridge, UK, AMP-0XG, AMP-0YR, AMP-11G, AMP-19R) were separately denatured in 70% formamide at 75 °C for 5 min, then slides were dehydrated through an ethanol series. Probes were cohybridised in pairs of 4 µL each of: chrX and chrY; chrX and chr19; chr11 and chr19. The probes were added to the slides, coverslips were sealed with rubber cement, and the slides were hybridised for 48 h at 37 °C. Coverslips were removed, and slides were washed in 0.7 × SSC, 0.3% Tween-20 at 73 °C for 3 min to remove unbound probe, then washed in 2 × SSC for 2 min at RT, rinsed in water and air-dried in the dark. Slides were counterstained with 16 µL VectorShield with DAPI (Vector Labs, Peterborough, UK) under a 22 × 50 mm cover slip and imaged at 100× on an Olympus BX-61 epifluorescence microscope equipped with a Hamamatsu Orca-ER C4742-80 cooled CCD camera and appropriate filters. Images were captured using Smart-Capture 3 (Digital Scientific UK, Cambridge, UK) with fixed exposure times for each fluorochrome.

2.3. Image Analysis

Analysis was performed using our image analysis software (Nuclear Morphology Analysis, available from http://bitbucket.org/bmskinner/nuclear_morphology/wiki/Home/, version 1.15.0) for morphometric analysis of mouse sperm shape [18]. Here, we combine nuclear morphometry with FISH signal detection in order to rigorously quantify the distribution of chromosome territories within the asymmetric mouse sperm head. Within our images we detected 1445 PWK nuclei, 906 LEWES nuclei and 712 STF nuclei across all hybridisations (Figure 1B). The number of nuclei with FISH signals detected which were used for chromosome positioning analysis are given in Table S1.

This analysis, which we refer to as nuclear cartography is a form of mesh warping, achieved by overlaying a mesh onto each individual sperm nucleus and quantifying the distribution of the chromosomal signal within each face of the mesh (Figure 1C). This allows accurate, quantifiable 2D analysis of the signal distribution in each cell. Subsequently, since the mesh overlaid onto each sperm head is structurally equivalent, dynamic image warping is used to combine multiple individual nuclear outlines onto the consensus shape of the cell population (Figure 1D). Using this method, signal intensity can be averaged over multiple sperm heads, reducing the effect of background inhomogeneities and revealing the consensus two-dimensional location of the signal in the population as a whole.

For successful warping of the source image, the face of the mesh to which each pixel belongs must be determined. The critical step is the construction of the mesh, such that each face contains a structurally equivalent region of the nucleus. First, we identify key landmarks around the periphery of the nucleus (i.e., the apical hook, tail attachment site, and other areas of maximal curvature), as described previously [18]. Next, semi-landmarks are constructed by spacing a set number of equidistant points between each landmark (Figure 1C-i). These then serve as the peripheral vertices of the mesh. The internal vertices are created by walking through the points pairwise from the tip of the nucleus, and generating a vertex at the centre of the line connecting each pair (Figure 1C-ii). Internal and peripheral vertices are connected into the faces of the mesh (Figure 1C-iii). The same structural mesh is created for the consensus nucleus shape, and for each individual nucleus. An affine transform is applied to image pixels within each face, moving them to their equivalent positions in the consensus mesh. After pixels have been relocated, a gap-filling kernel sets any empty pixel to the average of the surrounding non-zero 8-connected pixels, as long as there are at least 4 non-zero surrounding pixels. This reduces smearing in cases where there is a large size difference between source and consensus mesh faces.

In this way, we warp the original images to fit the consensus nucleus. The warped images can be combined to reveal the locations of consistent nuclear signal. Random noise is averaged out, while consistent signals are reinforced. To avoid bias from higher or lower intensity signals in different nuclei, the FISH images are binarised before warping. Since the individual images are being warped to fit a template shape, it is possible to choose any template with the same underlying graph structure in the mesh. This allows comparison of FISH signal distributions between different hybridisations.

To compare signal distributions between warped signals, we used an open source implementation of a multi-scale structural similarity index measure (MS-SSIM*) [19,20], which quantifies visual similarity between images [21] on a scale of 0 (no similarity) to 1 (identical images). To further assess adjacency of chromosome territories, we identified the chromosomal signals within the nuclei by thresholding [3], and measured the distances between the centres of mass of co-hybridised chromosomes. Statistical analyses were performed in R 3.5.1 [22], and charts were generated using the cividis colour palette [23].

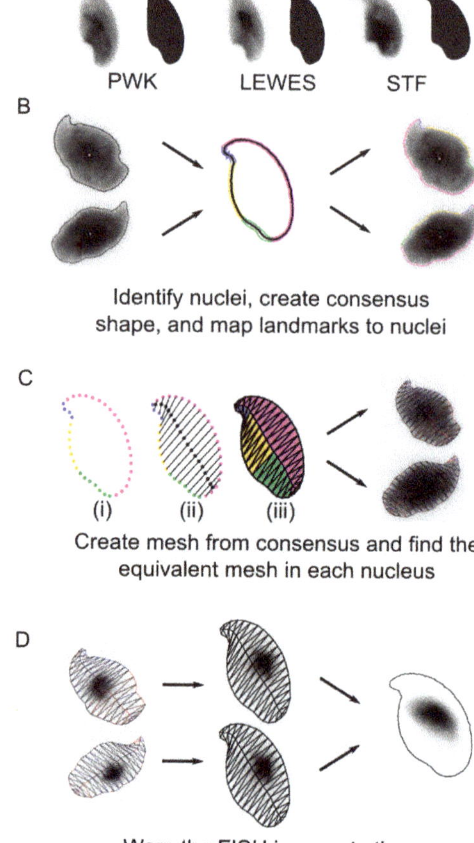

Figure 1. The process of warping fluorescence in-situ hybridisation (FISH) images. (**A**) Examples of un-FISHed nuclei from the three strains, as described in [18]. (**B**) After FISH, nuclei are automatically identified and landmarks are discovered. (**C**) A mesh is created from the consensus nuclear shape; (i) peripheral vertices are evenly spaced between landmarks; (ii) internal vertices divide vertex pairs from the tip; (iii) all vertices are joined. The equivalent mesh is constructed for each nucleus. (**D**) The FISH signal image is transformed to move every pixel to its location in the consensus mesh. The warped images are combined to yield the composite signal image. Mouse strains *Mus musculus musculus* (PWK), *M. m. domesticus* (LEWES) and *Mus spretus* (STF).

3. Results

3.1. The Sex Chromosomes Have Conserved Position in Mouse Sperm Nuclei

The process of hybridising FISH probes to sperm nuclei required a considerable swelling step due to the highly compact chromatin. The nuclear area doubles from about 20 μm^2 to about 40 μm^2, with the majority of the swelling in the dorsal/ventral axis (Figure S3). This swelling distorts the nuclear shape; our method for automated nucleus and landmark detection [18] was able to identify and orient swelled nuclei successfully, despite the fewer landmarks available.

Confident that we could orient a FISH signal within the nucleus, we applied the new technique to FISH images of mouse sperm from three strains, using chromosome paints for the X and Y chromosomes. These have been previously reported in *M. musculus* strain C57Bl6 to lie under the acrosome [8,9]. Nuclei and signals were detected from the captured images, a consensus nuclear shape was calculated for each strain, and each FISH image was warped onto that consensus shape. A composite image was created by layering each FISH image, effectively providing a heat-map of signal location within the nucleus.

Our results confirm a consistent sub-acrosomal location for both X and Y chromosomes (Figure 2). Following the signal warping onto the population consensus, we observed that both X and Y chromosomes have overlapping territories (Figures 3 and 4).

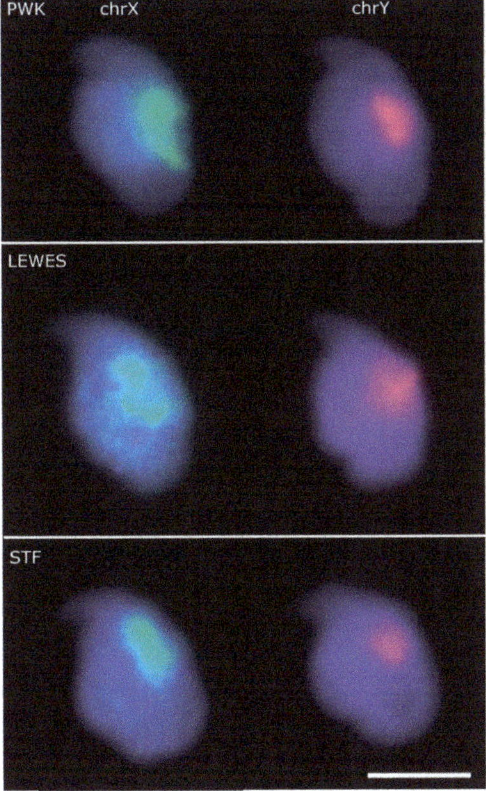

Figure 2. Example images showing the sex chromosome positions within the three strains. Scale bar represents 5 μm.

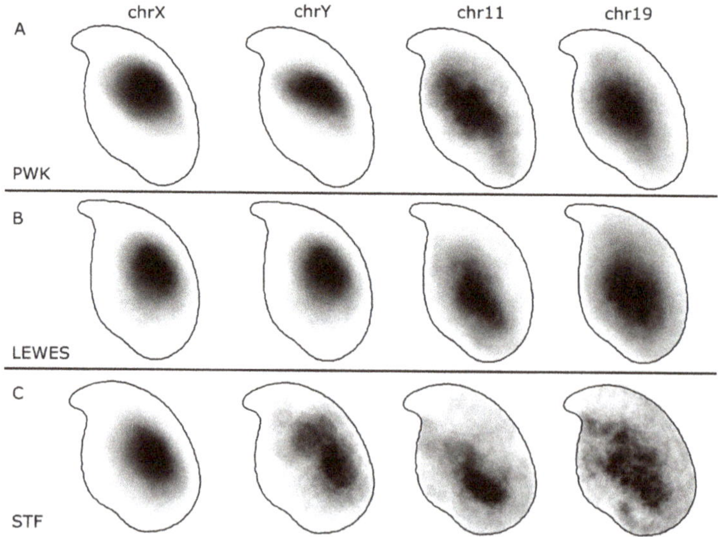

Figure 3. Composite signal distributions for chromosomes X, Y, 11 and 19 in (**A**) PWK, (**B**) LEWES and (**C**) STF. The sex chromosomes occupy a consistent territory apical and dorsal to the centre of mass, generally under the acrosome but rarely extending fully to the periphery of the nucleus. Chromosomes 11 and 19 are more widely distributed, with the predominant location basal and ventral to the centre of mass.

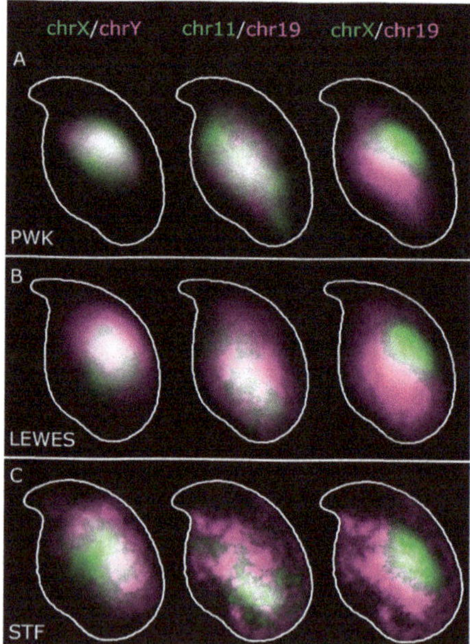

Figure 4. Overlay of warped distributions from Figure 3 shows the similarities between chromosome X and Y territories, and 11 and 19 territories in (**A**) *PWK*; (**B**) *LEWES*; and (**C**) *STF*. White shows regions of overlap. Chromosomes X and 19 (and X and 11) are predominantly non-overlapping.

3.2. Chromosomes 11 and 19 Occupy Similar Nuclear Addresses

With the sex chromosome locations confirmed to be conserved, we decided to examine two further chromosomes, both of which have previously been reported in the literature. Chromosome 19 has been described in C57Bl/6 mice to frequently lie toward the base of the nucleus [8]. Furthermore in Hi-C experiments, chromosomes X and 19 had a low association in *M. musculus* C57BL sperm chromatin; chromosome 19 and chromosome 11 had a moderate association with each other [17]. For this reason, we hypothesised that chr11 and chr19 might share a similar distribution, and that this would be distinct from that of the sex chromosomes.

The composite signal position data are shown in Figure 3. The patterns are indeed different to that of the sex chromosomes. The majority of the signal lies ventral and basal to the centre of the nucleus, yet there are clearly instances of signal throughout the nucleus, from the basal region near the tail attachment point to the apex and partially extending into the hook. Some examples of these positions in individual nuclei are shown in Figure 5.

Although hybridization efficiency was poorer in *M. spretus*, the same patterns are apparent as in the *M. musculus* strains. Interestingly, we observed instances of both chr11 and chr19 below the acrosomal curve, in which the chr19 was generally more elongated than chr11 (see Figure 5B,F). Where chromosome 19 was co-hybridised with chromosome X, we were able to see rare instances of chrX and chr19 lying adjacent, with chrX more internal (Figure S1).

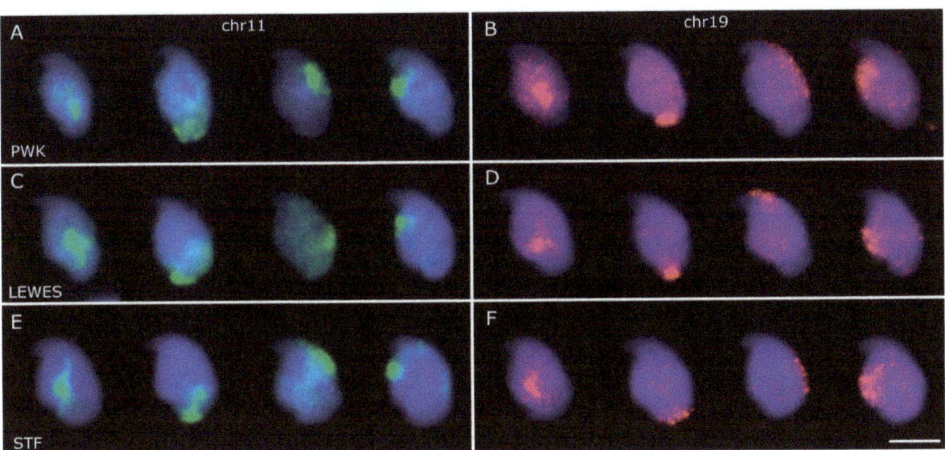

Figure 5. Examples of individual chromosome positions for chr11 (**A,C,E**) and chr19 (**B,D,F**) in the three strains; the chr11 and chr19 panels do not show the same nuclei. While the majority of the signals for each chromosome were observed ventral and basal of the nuclear centre (column 1), we found territories at the base of the nucleus (column 2), under the acrosome (column 3), and along the ventral surface below the hook (column 4). Scale bar represents 5 μm.

Given the similarity in overall signal distributions, we looked to see if chr11 and chr19 tend to lie adjacent to each other in individual nuclei. Visually, we can see that they are occasionally adjacent, but are not always associated. Measurement of the distance between the chromosome signal centers of mass showed no difference between chr11 and 19 or between chr11 and X, nor did a comparison of individual nucleus warped signal images via a MS-SSIM*, a technique also used in comparisons of radiological images [24] ($p > 0.05$, Wilcoxon rank sum tests; Figure 6). We conclude that, although chr11 and chr19 have a similar range of possible addresses to occupy within an individual sperm head, they do not necessarily interact, and are no more likely to be adjacent than chromosomes 11 and X. It is however important to appreciate that our data addresses chromosome territories as a whole, rather than individual loci, and further work will be needed to robustly compare our data with the Hi-C data from [17] (see also Section 4).

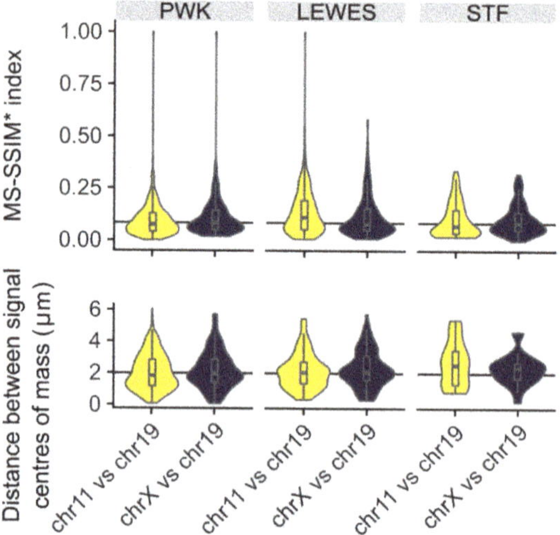

Figure 6. Chromosomes 11 and 19 do not colocalize within individual nuclei; colocalization of signals shows no difference comparing chr11 and chr19 as when comparing chrX and chr19 by either multi-scale structural similarity index (MS-SSIM*) (upper) or the distances between the chromosome signal centers (lower).

3.3. Quantification of Signal Positions Reveals Conserved Chromosome Organisation across Species

In order to quantify the similarity of signal locations both within and between strains, we warped images from all three strains onto the LEWES (domesticus) consensus outline. These consensus warped images were compared using MS-SSIM*, revealing the similarities in the range of possible nuclear addresses a chromosome could occupy in each strain. The X and Y territories had high structural similarity to each other in all three strains, and had high concordance between strains (Figure 7). Similarly, we saw greater similarity between chr11 and chr19 in all three strains. The pattern was slightly less clear between *M. spretus* and the other strains, presumably due to the lower hybridisation efficiency of the probes. To confirm there was no artefactual bias introduced by the choice of LEWES as the destination shape, we examined the effect of warping signals onto either the PWK or STF consensus outlines, and found that this made little difference in the values obtained (see also Figure S2, Table S2). This demonstrates that our method is robust for comparing differently shaped nuclei as long as we can define structurally equivalent landmarks.

4. Discussion

We have presented here a new method for quickly and efficiently mapping chromosome position in asymmetric nuclei, such as sperm, based on linking chromosome signals with morphometric information about nuclear structure. Using this analysis, we have been able to measure and quantify differences in chromosome territory position in sperm from three mouse strains. All mouse strains studied here diverged, at most, 3 million years ago [25,26], and the karyotypes of *M. musculus* and *M. spretus* both have 40 chromosomes [27]. *M. musculus* and *M. spretus* are able to produce hybrids in laboratory conditions, of which the female F1 is fertile [28]. We have demonstrated here that orthologous chromosomes adopt similar conformations in the three strains, despite differences in nuclear shape.

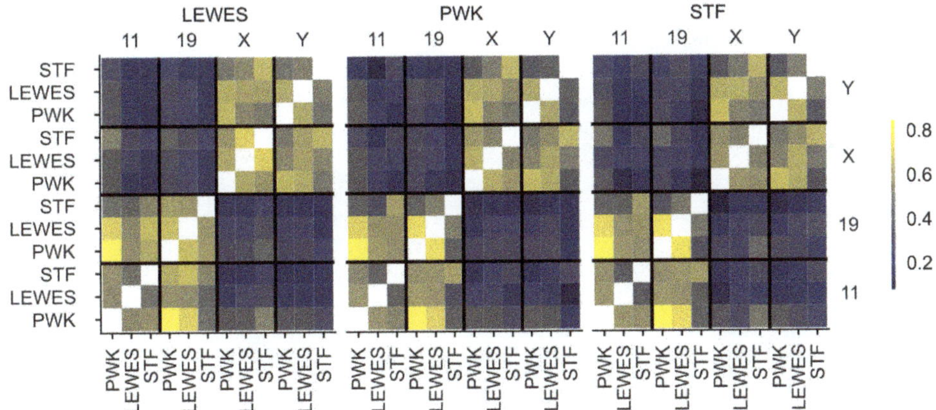

Figure 7. Similarity of signal distributions in composite warped images measured by MS-SSIM*, on a scale of 0–1, where 0 indicates no similarity, and 1 indicates identical images. Images were warped in turn onto the consensus shapes of LEWES, PWK and STF. There is high correlation between the MS-SSIM* scores obtained when images are warped onto different target shapes (see Figure S2). Both within strains and between strains, there is a clear similarity between the distributions of chrX and chrY, and chr11 and chr19, but little similarity between the reciprocal combinations.

4.1. Chromosomes X and Y Have a Conserved Dorsal/Sub-Acrosomal Position

Both the mouse X and Y chromosomes have been subject to massive amplification of euchromatic sequences. The full sequence of a *M. m. musculus* C57Bl/6 Y chromosome revealed the complex ampliconic structure [29], and demonstrated the presence of similar amplicons on the *M. spretus* Y. These amplicons are thought to arise from genomic conflict in spermatids [30], and copy number measurements of individual ampliconic genes suggests *M. spretus* has generally amplified the same gene families as *M. musculus*, with the exception of X-linked H2al1, which has amplified specifically in the *M. musculus* lineage.

Despite the close evolutionary relationship of *M. musculus* and *M. spretus*, some small rearrangements involving the sex chromosomes have been documented—for example, the Clcn4 gene, X-linked in most mammals including *M. spretus*, is autosomal in *M. musculus* [31], with clear translocation breakpoints surrounding the gene [32].

Given the overall structural similarity of the orthologous chromosomes, it is likely they occupy a similar volume within the nucleus, and are subject to similar conformational constraints. The sex chromosomes have been previously described to adopt a dorsal position in the rodent sperm nucleus [8,9], and have been seen to be sub-acrosomal in human, marsupial and monotreme sperm [14]. It has been suggested that the X chromosome in X-bearing sperm is the first to enter the egg during fertilisation. The position of the Y in marsupials is not reported, but as in mice, it is likely that the Y adopts the same position as X simply because the space is available. In monotremes, the platypus Y chromosomes do show a similar distribution to the X chromosomes [33]. Since the sex chromosomes are different sizes—approximately 90 Mb versus 170 Mb—there must be differences in the chromatin packing to allow them to occupy the same nuclear volume. In future we will be interested to study the impact of chromosome constitution on nuclear morphology.

4.2. Chromosomes 11 and 19 Have a Conserved Ventral/Basal Distribution

Chromosome 19 has been observed by others to lie in the basal region of the nucleus in approximately two thirds of nuclei based on imaging and manually scoring at least 350 *M. musculus* C57Bl/6 sperm nuclei [8,9]. Our results support these data, and demonstrate conservation of this position across species. The signal in *M. spretus* is less clear, likely due to the cross-species hybridisation, but the pattern is still distinguishable.

Our data from co-hybridisations suggest that although chr11 and chr19 adopt a similar overall location, they do not always lie adjacent within a single nucleus. This indicates that while they have preferred regions of the nucleus, they are mostly unconstrained with regard to each other. Aggregate data from Hi-C experiments in C57Bl/6 sperm [17] have indicated that chr19 is infrequently associated with the X chromosome (and by inference, the Y chromosome), and that chr11 is only moderately associated with both chrX and chr19. It is, however, currently unclear why Hi-C shows chromosome 19 to be more strongly associated with chromosome 11 than the X chromosome, given our data showing that these three chromosome territories are on average equidistant. One potential explanation is that while our measurements focus on the centre of each chromosome territory, interactions occur at the periphery of territories in cells where they abut each other. The mouse sperm head tends to have a DAPI-dense chromocenter core, and that the X/Y and 11/19 regions are deduced to usually lie on opposite sides of this. Potentially, this core forms a barrier to inter-chromosomal interactions (Figure 8). As an analogy, Cersei and Jaime (chromosomes 11 and 19) may both live in the ground floor flat, but they do not take up the exact same physical space, remaining on average a few meters apart. Meanwhile, their upstairs neighbor Daenerys (chromosome X or Y) is roughly equidistant from them, but does not interact with them due to the barrier in between (the centric heterochromatin). However, when averaged across the course of many days, Cersei and Jaime collectively occupy the downstairs flat, while Daenerys occupies the spatially distinct upper floor. A higher resolution investigation of individual loci found to be associated in the Hi-C data will help resolve this distribution.

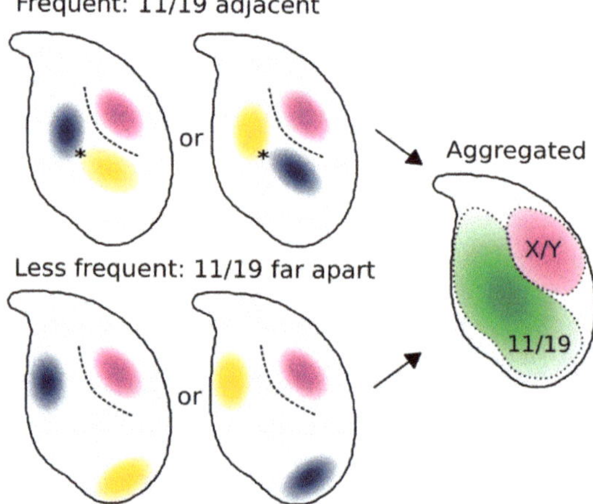

Figure 8. A simple model of how our data may relate individual cells to aggregate measurements. In individual cells, chr11 and chr19 (blue/yellow) frequently lie adjacent, and more rarely further apart. Chromosomes X and Y (purple) lie consistently below the acrosome. In contrast, chromosomes 11 and 19 do not have strictly fixed addresses, but reside interchangeably within the same general area of the nucleus. Thus, chromosomes 11 and 19 colocalise in the aggregate distribution despite not overlapping within any individual nucleus. In this model, the chromocenter core acts as a physical barrier to interchromosomal interactions, explaining why Hi-C detects more 11/19 interactions (indicated by *) than 11/X or 19/X interactions despite the similar physical distances between the centres of mass of the three territories.

Overall, our measurements tend to support previous Hi-C and FISH findings in laboratory mouse sperm, and provide evidence that the same patterns will be found in *M. spretus*. The concept of spatial synteny—the conserved 3D position of orthologous loci despite karyotypic rearrangements—has been proposed [34], and there is increasing evidence for conserved nuclear organization of genes following chromosomal rearrangements [35]. As we extend our studies, it will be interesting to compare the positions of the full set of chromosomes, to better understand how the shorter and fatter *M. spretus* nucleus maps on the longer, thinner *M. musculus* nucleus. Further comparisons with other mouse strains with greater shape variability will also be of value; for example BALB/c, which have frequent shape abnormalities and aneuploidies [18,36].

Studies of strains with other aneuploidies, chromosomal rearrangements or Robertsonian fusions, which will additionally constrain chromosome territories will be of interest. In humans, no gross morphological differences in sperm nuclei have been seen in men carrying Robertsonian fusions [37]. However, in boars (*Sus scrofa*), while gross nuclear morphology was not perturbed in animals carrying a t(13;17) Robertsonian translocation, differences were apparent in the positions of the affected chromosomes [38]. Extending beyond mice, rats (*Rattus rattus*) have a much thinner hooked sperm nucleus; rat chromosomes have been mapped in developing spermatids from stages 7–13. The nucleus is compressed from a structure which at stage 10 is markedly similar to a mature mouse sperm nucleus [39]. The associated dynamics of nuclear reshaping during spermiogenesis, and chromosome repositioning are an area of active research [10].

4.3. This Method Allows Rapid Screening of Large Numbers of Nuclei

In this analysis, we examined more than 3000 nuclei, and the method scales to greater numbers with little additional time or user effort after images have been captured. The warping algorithm processed these nuclei in under half an hour on a desktop computer equipped with an Intel i5-2400 processor and 16 Gb memory, with the total user time excluding image capture being a few hours. This is of course experience and hardware dependent, but the key point is that the total analysis time can be measured in hours rather than days. Importantly, our analysis does not rely on extensive manual classification of chromosome position, making it less subjective than current approaches, and amenable to automation. The use of a mesh to warp signals from different nuclei onto a single template shape allows for quantitative measurements of the similarity of signal distributions between images, and in principle will allow us to study small differences in locus position that have been beyond the scope of current scoring systems. Beyond chromosome territory positioning, it is also amenable to the study of single BAC probes, or any small probe generating a punctuate signal, as long as sufficient nuclei are analyzed to generate an aggregate signal; together with Hi-C data this will allow us to study which intra- and inter-chromosomal folding contacts are retained in the sperm head, and address long standing questions of whether sperm chromatin organisation represents an echo of round spermatid chromatin organisation, or prefigures future chromatin folding dynamics in the fertilised zygote.

A further methodological interest would be to identify reliable internal structural features within the nucleus, using DAPI or other stains. Currently we use only peripheral features as landmarks, which puts limits on the accuracy of our mesh when deforming images. More internal structural data would permit more complex morphometric approaches such as Teichmüller mapping, which has been used successfully in analysis (for example) of wing shape in Drosophila species [40].

5. Conclusions

Here we have demonstrated a new method for locating chromosome paints or other nuclear signals within mouse sperm nuclei, which is in principle also applicable to other asymmetric nuclei, including nuclei with fewer axes of asymmetry, such as spatulate sperm nuclei. We have used this technique to confirm the non-random positioning of the sex chromosomes, and of chromosomes 11 and 19, and demonstrated quantitation of signal positions allowing comparison between different strains and species. Importantly, we have integrated this method into existing open-source image analysis software designed for other biologists.

Supplementary Materials: The following are available online at http://www.mdpi.com/2073-4425/10/2/109/s1, Figure S1: Chromosomes X and 19 co-hybridization; Figure S2: Comparison of MS-SSIM * scores using different warping templates; Figure S3: Examples of swelled and unswelled nuclei; Table S1: Numbers of nuclei with FISH signals analysed in this study., Table S2: Complete MS-SSIM* comparisons between warped composite images, including the individual similarity components of contrast, luminance and structure.

Author Contributions: Conceptualization, B.M.S. and P.E.; Methodology, B.M.S. and P.E.; Software and Validation, B.M.S.; Investigation, J.B., C.C.R.; Data Curation and Formal Analysis, B.M.S.; Visualization, B.M.S.; Supervision and Project Administration, P.E.; Writing—Original Draft, B.M.S. and P.E.; Writing—Review and Editing, B.M.S., C.C.R., J.M.G., E.L.L. and P.E.; Resources, J.M.G., E.L.L., E.E.K.K., N.A. and P.E.; Funding Acquisition, N.A. and P.E.

Funding: B.M.S. was supported by the Biotechnology and Biological Sciences Research Council (BBSRC, BB/N000129/1). P.E. and C.C.R. were supported by H.E.F.C.E. (University of Kent) and by the BBSRC (BB/N000463/1). J.M.G. and E.L.L. were supported by the Eunice Kennedy Shriver National Institute of Child Health and Human Development of the National Institutes of Health (R01-HD073439 and R01-HD094787) and the National Institute of General Medical Sciences (R01-GM098536). E.E.K.K. was supported by the National Science Foundation Graduate Research Fellowship Program under Grant No. (DGE-1313190).

Acknowledgments: We thank the animal handling staff at the University of Montana.

Conflicts of Interest: The authors declare no conflict of interest. The funders had no role in the design of the study; in the collection, analyses, or interpretation of data; in the writing of the manuscript, or in the decision to publish the results.

References

1. Skinner, B.M.; Volker, M.; Ellis, M.; Griffin, D.K. An appraisal of nuclear organisation in interphase embryonic fibroblasts of chicken, turkey and duck. *Cytogenet. Genome Res.* **2009**, *126*, 156–164. [CrossRef] [PubMed]

2. Foster, H.A.; Griffin, D.K.; Bridger, J.M. Interphase chromosome positioning in in vitro porcine cells and ex vivo porcine tissues. *BMC Cell Biol.* **2012**, *13*, 30. [CrossRef] [PubMed]

3. O'Connor, R.E.; Kiazim, L.; Skinner, B.; Fonseka, G.; Joseph, S.; Jennings, R.; Larkin, D.M.; Griffin, D.K. Patterns of microchromosome organization remain highly conserved throughout avian evolution. *Chromosoma* **2018**. [CrossRef] [PubMed]

4. Foster, H.A.; Abeydeera, L.R.; Griffin, D.K.; Bridger, J.M. Non-random chromosome positioning in mammalian sperm nuclei, with migration of the sex chromosomes during late spermatogenesis. *J. Cell Sci.* **2005**, *118*, 1811–1820. [CrossRef] [PubMed]

5. Millan, N.M.; Lau, P.; Hann, M.; Ioannou, D.; Hoffman, D.; Barrionuevo, M.; Maxson, W.; Ory, S.; Tempest, H.G. Hierarchical radial and polar organisation of chromosomes in human sperm. *Chromosome Res.* **2012**, *20*, 875–887. [CrossRef] [PubMed]

6. Ioannou, D.; Millan, N.M.; Jordan, E.; Tempest, H.G. A new model of sperm nuclear architecture following assessment of the organization of centromeres and telomeres in three-dimensions. *Sci. Rep.* **2017**, *7*, 41585. [CrossRef] [PubMed]

7. Zalenskaya, I.A.; Zalensky, A.O. Non-random positioning of chromosomes in human sperm nuclei. *Chromosome Res.* **2004**, *12*, 163–173. [CrossRef]

8. Kocer, A.; Henry-Berger, J.; Noblanc, A.; Champroux, A.; Pogorelcnik, R.; Guiton, R.; Janny, L.; Pons-Rejraji, H.; Saez, F.; Johnson, G.D.; et al. Oxidative DNA damage in mouse sperm chromosomes: Size matters. *Free Radic. Biol. Med.* **2015**, *89*, 993–1002. [CrossRef]

9. Champroux, A.; Damon-Soubeyrand, C.; Goubely, C.; Bravard, S.; Henry-Berger, J.; Guiton, R.; Saez, F.; Drevet, J.; Kocer, A. Nuclear Integrity but not topology of mouse sperm chromosome is affected by oxidative DNA damage. *Genes* **2018**, *9*, 501. [CrossRef]

10. Namekawa, S.H.; Park, P.J.; Zhang, L.-F.; Shima, J.E.; McCarrey, J.R.; Griswold, M.D.; Lee, J.T. Postmeiotic sex chromatin in the male germline of mice. *Curr. Biol.* **2006**, *16*, 660–667. [CrossRef]

11. Finch, K.A.; Fonseka, K.G.L.; Abogrein, A.; Ioannou, D.; Handyside, A.H.; Thornhill, A.R.; Hickson, N.; Griffin, D.K. Nuclear organization in human sperm: Preliminary evidence for altered sex chromosome centromere position in infertile males. *Hum. Reprod.* **2008**, *23*, 1263–1270. [CrossRef]

12. Olszewska, M.; Wiland, E.; Kurpisz, M. Positioning of chromosome 15, 18, X and Y centromeres in sperm cells of fertile individuals and infertile patients with increased level of aneuploidy. *Chromosome Res.* **2008**, *16*, 875–890. [CrossRef] [PubMed]

13. Ioannou, D.; Griffin, D.K. Male fertility, chromosome abnormalities, and nuclear organization. *Cytogenet. Genome Res.* **2010**, *133*, 269–279. [CrossRef] [PubMed]

14. Greaves, I.K.; Rens, W.; Ferguson-Smith, M.A.; Griffin, D.; Marshall Graves, J.A. Conservation of chromosome arrangement and position of the X in mammalian sperm suggests functional significance. *Chromosome Res.* **2003**, *11*, 503–512. [CrossRef] [PubMed]

15. Stevens, T.J.; Lando, D.; Basu, S.; Atkinson, L.P.; Cao, Y.; Lee, S.F.; Leeb, M.; Wohlfahrt, K.J.; Boucher, W.; O'Shaughnessy-Kirwan, A.; et al. 3D structures of individual mammalian genomes studied by single-cell Hi-C. *Nature* **2017**, *544*, 59. [CrossRef] [PubMed]

16. Nagano, T.; Lubling, Y.; Stevens, T.J.; Schoenfelder, S.; Yaffe, E.; Dean, W.; Laue, E.D.; Tanay, A.; Fraser, P. Single-cell Hi-C reveals cell-to-cell variability in chromosome structure. *Nature* **2013**, *502*, 59. [CrossRef]

17. Battulin, N.; Fishman, V.S.; Mazur, A.M.; Pomaznoy, M.; Khabarova, A.A.; Afonnikov, D.A.; Prokhortchouk, E.B.; Serov, O.L. Comparison of the three-dimensional organization of sperm and fibroblast genomes using the Hi-C approach. *Genome Biol.* **2015**, *16*, 77. [CrossRef]

18. Skinner, B.M.; Rathje, C.C.; Bacon, J.; Johnson, E.E.P.; Larson, E.L.; Kopania, E.E.K.; Good, J.M.; Yousafzai, G.; Affara, N.A.; Ellis, P.J.I. A high-throughput method for unbiased quantitation and categorisation of nuclear morphology. *bioRxiv* **2018**, 312470. [CrossRef]

19. Prieto, G.; Chevalier, M.; Guibelalde, E. *MS_SSIM Index and MS_SSIM* Index as a Java Plugin for ImageJ*; Department of Radiology, Faculty of Medicine, Universidad Complutense: Madrid, Spain, 2009.

20. Rouse, D.M.; Hemami, S.S. Analyzing the role of visual structure in the recognition of natural image content with multi-scale SSIM. In Proceedings of the Human Vision and Electronic Imaging XIII, San Jose, CA, USA, 27–31 January 2008; International Society for Optics and Photonics: Bellingham, WA, USA, 2008; Volume 6806, p. 680615.

21. Wang, Z.; Bovik, A.C.; Sheikh, H.R.; Simoncelli, E.P. Image quality assessment: From error visibility to structural similarity. *IEEE Trans. Image Process.* **2004**, *13*, 600–612. [CrossRef]

22. R Core Team R. *A Language and Environment for Statistical Computing*; R Foundation for Statistical Computing: Vienna, Austria, 2018.

23. Nuñez, J.R.; Anderton, C.R.; Renslow, R.S. Optimizing colormaps with consideration for color vision deficiency to enable accurate interpretation of scientific data. *PLoS ONE* **2018**, *13*, e0199239. [CrossRef]

24. Renieblas, G.P.; Nogués, A.T.; González, A.M.; Gómez-Leon, N.; del Castillo, E.G. Structural similarity index family for image quality assessment in radiological images. *J. Med. Imaging* **2017**, *4*, 035501. [CrossRef] [PubMed]

25. Fabre, P.-H.; Hautier, L.; Dimitrov, D.; Douzery, E.J. A glimpse on the pattern of rodent diversification: A phylogenetic approach. *BMC Evol. Biol.* **2012**, *12*, 88. [CrossRef] [PubMed]

26. Chevret, P.; Veyrunes, F.; Britton-Davidian, J. Molecular phylogeny of the genus *Mus* (Rodentia: Murinae) based on mitochondrial and nuclear data. *Biol. J. Linn. Soc.* **2005**, *84*, 417–427. [CrossRef]

27. Silver, L.M. *Mouse Genetics: Concepts and Applications*; Oxford University Press: Oxford, NY, USA, 1995; ISBN 978-0-19-507554-0.

28. Palomo, L.J.; Justo, E.R.; Vargas, J.M. *Mus spretus* (Rodentia: Muridae). *Mamm. Species* **2009**, *840*, 1–10. [CrossRef]

29. Soh, Y.Q.S.; Alföldi, J.; Pyntikova, T.; Brown, L.G.; Graves, T.; Minx, P.J.; Fulton, R.S.; Kremitzki, C.; Koutseva, N.; Mueller, J.L.; et al. Sequencing the mouse Y chromosome reveals convergent gene acquisition and amplification on both sex chromosomes. *Cell* **2014**, *159*, 800–813. [CrossRef] [PubMed]

30. Ellis, P.J.I.; Bacon, J.; Affara, N.A. Association of *Sly* with sex-linked gene amplification during mouse evolution: A side effect of genomic conflict in spermatids? *Hum. Mol. Genet.* **2011**, *20*, 3010–3021.

31. Rugarli, E.I.; Adler, D.A.; Borsani, G.; Tsuchiya, K.; Franco, B.; Hauge, X.; Disteche, C.; Chapman, V.; Ballabio, A. Different chromosomal localization of the *Clcn4* gene in *Mus spretus* and C57BL/6J mice. *Nat. Genet.* **1995**, *10*, 466–471. [CrossRef]

32. Nguyen, D.K.; Yang, F.; Kaul, R.; Alkan, C.; Antonellis, A.; Friery, K.F.; Zhu, B.; de Jong, P.J.; Disteche, C.M. *Clcn4-2* genomic structure differs between the X locus in *Mus spretus* and the autosomal locus in *Mus musculus*: AT motif enrichment on the X. *Genome Res.* **2011**, *21*, 402–409. [CrossRef]

33. Tsend-Ayush, E.; Dodge, N.; Mohr, J.; Casey, A.; Himmelbauer, H.; Kremitzki, C.L.; Schatzkamer, K.; Graves, T.; Warren, W.C.; Grützner, F. Higher-order genome organization in platypus and chicken sperm and repositioning of sex chromosomes during mammalian evolution. *Chromosoma* **2009**, *118*, 53–69. [CrossRef]

34. Veron, A.; Lemaitre, C.; Gautier, C.; Lacroix, V.; Sagot, M.-F. Close 3D proximity of evolutionary breakpoints argues for the notion of spatial synteny. *BMC Genom.* **2011**, *12*, 303. [CrossRef]

35. Dai, Z.; Xiong, Y.; Dai, X. Neighboring genes show interchromosomal colocalization after their separation. *Mol. Biol. Evol.* **2014**, *31*, 1166–1172. [CrossRef]

36. Kishikawa, H.; Tateno, H.; Yanagimachi, R. Chromosome analysis of BALB/c mouse spermatozoa with normal and abnormal head morphology. *Biol. Reprod.* **1999**, *61*, 809–812. [CrossRef] [PubMed]

37. Cassuto, N.G.; Le Foll, N.; Chantot-Bastaraud, S.; Balet, R.; Bouret, D.; Rouen, A.; Bhouri, R.; Hyon, C.; Siffroi, J.P. Sperm fluorescence in situ hybridization study in nine men carrying a Robertsonian or a reciprocal translocation: Relationship between segregation modes and high-magnification sperm morphology examination. *Fertil. Steril.* **2011**, *96*, 826–832. [CrossRef] [PubMed]

38. Acloque, H.; Bonnet-Garnier, A.; Mompart, F.; Pinton, A.; Yerle-Bouissou, M. Sperm Nuclear architecture is locally modified in presence of a Robertsonian Translocation t(13;17). *PLoS ONE* **2013**, *8*, e78005. [CrossRef] [PubMed]

39. Meyer-Ficca, M.; Muller-Navia, J.; Scherthan, H. Clustering of pericentromeres initiates in step 9 of spermiogenesis of the rat (*Rattus norvegicus*) and contributes to a well defined genome architecture in the sperm nucleus. *J. Cell Sci.* **1998**, *111*, 1363–1370. [PubMed]

40. Choi, G.P.T.; Mahadevan, L. Planar morphometrics using Teichmüller maps. *Proc. R. Soc. A Math. Phys. Eng. Sci.* **2018**, *474*, 20170905. [CrossRef] [PubMed]

Article

Nucleolar Expression and Chromosomal Associations in Robertsonian Spermatocytes of *Mus musculus domesticus*

Fernanda López-Moncada [1], Daniel Tapia [1], Nolberto Zuñiga [1], Eliana Ayarza [2], Julio López-Fenner [3], Carlo Alberto Redi [4] and Soledad Berríos [1,*]

1   Programa Genética Humana, ICBM, Facultad de Medicina, Universidad de Chile, Santiago 8380453, Chile; fernanda.lopez.m@gmail.com (F.L.-M.); danieltapiasalvo@gmail.com (D.T.); nolbertoz@ug.uchile.cl (N.Z.)
2   Departamento de Tecnología Médica, Facultad de Medicina, Universidad de Chile, Santiago 8380453, Chile; elianaayarza@yahoo.com
3   Departamento de Ingeniería Matemática, Universidad de La Frontera, Temuco, Chile; julio.lopez@ufrontera.cl
4   Department of Biology and Biotechnology "L. Spallanzani", University of Pavia, 27100 Pavia, Italy; carloalberto.redi@unipv.it
*   Correspondence: sberrios@med.uchile.cl

Received: 17 December 2018; Accepted: 5 February 2019; Published: 6 February 2019

Abstract: We studied and compared the nucleolar expression or nucleoli from specific bivalents in spermatocytes of the standard *Mus musculus domesticus* 2n = 40, of Robertsonian (Rb) homozygotes 2n = 24 and heterozygotes 2n = 32. We analyzed 200 nuclear microspreads of each specific nucleolar chromosome and spermatocyte karyotype, using FISH to identify specific nucleolar bivalents, immunofluorescence for both fibrillarin of the nucleolus and the synaptonemal complex of the bivalents, and DAPI for heterochromatin. There was nucleolar expression in all the chromosomal conditions studied. By specific nucleolar bivalent, the quantitative relative nucleolar expression was higher in the bivalent 12 than in its derivatives, lower in the bivalent 15 than in its derivatives and higher in the bivalent 16 than its Rb derivatives. In the interactions between non-homologous chromosomal domains, the nucleolar bivalents were preferentially associated through pericentromeric heterochromatin with other bivalents of similar morphology and sometimes with other nucleolar bivalents. We suggest that the nucleolar expression in Rb nucleolar chromosomes is modified as a consequence of different localization of ribosomal genes (NOR) in the Rb chromosomes, its proximity to heterochromatin and its associations with chromosomes of the same morphology.

Keywords: nucleoli; NOR; chromosome associations; meiotic prophase; spermatocytes; *Mus m. domesticus*; Robertsonian chromosomes; chromosome translocation

1. Introduction

The nucleolar organizing regions (NORs) are chromosomal segments that contain repeated tandem sequences of ribosomal genes and can produce nucleoli [1–3]. In the species *Mus musculus domesticus* (2n = 40) these regions are multiple and are found in the telocentric chromosomes 12, 15, 16, 18 and 19 [4,5]. During the early stages of the meiotic prophase I, the ribosomal genes are actively transcribed, which makes it possible to observe nucleolar material bound to the NOR of the nucleolar bivalent that gives rise to it. The number of nucleoli observed during the meiotic prophase depends on the number of chromosomes with NOR, their transcriptional activity and the associations between nucleolar bivalents that can give rise to common nucleoli [6,7]. All the chromosomes of *M. m. domesticus* present large segments of heterochromatin in sectors adjacent to the centromeres and in the nucleolar chromosomes limiting the NORs [8,9]. Thus, in spermatocytes of *M. m. domesticus* 2n = 40, the suprachromosomal interactions that lead to the formation of nuclear territories would be especially determined by the association of groups of bivalents bound by their pericentromeric heterochromatin and the nucleoli would be part of some of those chromocentres [10]. *Mus m. domesticus* is characterized by natural populations of great heterogeneity in their diploid numbers due to the occurrence of Robertsonian (Rb) chromosome fusions [11,12]. In these chromosomal rearrangements, rupture at the centromere level occurs in two telocentric chromosomes and the subsequent fusion of their long arms, which generates Robertsonian metacentric chromosomes. These rearrangements occur in different combinations between all the telocentric chromosomes [12]. When the Rb fusions involve nucleolar chromosomes, these NORs are structurally preserved because the nucleolar organizing regions are located in the sub-centromeric region of the long arms of the chromosomes in this species [13]. In this way, in the nucleolar chromosomes of Rb mice, the NORs are located close to the centromeric region of a metacentric chromosome and surrounded by the pericentromeric heterochromatin from the two original ancestral telocentric chromosomes [14]. This new chromosomal organization modifies the distribution of NOR and nucleoli in the nucleus and the conformation of the territories in which they participate. In fact, spermatocytes of 2n = 40 mice, which present pericentromeric NORs exclusively in telocentric bivalents, show only nucleoli located in the nuclear periphery, while in Rb homozygous spermatocytes, nucleoli can be observed in the periphery and in the center of the nuclear space [15]. However, it is unknown if the change in the chromosomal position of the NOR (and therefore in the nuclear space) affects the magnitude of its expression in the meiotic prophase [16].

We studied whether the nucleolar expression varies comparing NORs localized in telocentric nucleolar chromosomes with the respective derived Rb chromosomes; as well as the interactions between these and other chromosomal domains in spermatocytes of 2n = 40 mice, homozygous for all telocentric chromosomes; 2n = 24, homozygous for Rb chromosomes; and 2n = 32, heterozygous for Rb chromosomes.

Nucleolar expression was observed in all the chromosomal conditions studied, being globally higher in the nuclei of spermatocytes 2n = 40 and 2n = 24 than in those from spermatocytes 2n = 32. By specific bivalent, the pattern of nucleolar expression was variable between telocentric nucleolar bivalents and it was modified in the derived Rb nucleolar chromosomes.

2. Materials and Methods

2.1. Animals

The spermatocytes of six three-months-old *M. m. domesticus* were analyzed. Two were homozygote 2n = 40 CD1 mice with all telocentric chromosomes and five pairs of nucleolar chromosomes numbers 12, 15, 16, 18 and 19. Two mice were Milano II 2n = 24 with three pairs of nucleolar Rb metacentric chromosomes, numbers 10.12, 5.15, 16.17, and two pairs of the telocentric nucleolar chromosomes, 18 and 19. The two heterozygotes Rb 2n = 32 had three single nucleolar Rb metacentric chromosomes, three single nucleolar telocentric chromosomes, numbers 12, 15, 16, and two pairs of telocentric nucleolar chromosomes, numbers 18 and 19. Chromosome numbers are reported according to the 2n = 40 standard karyotype [4,5]. The heterozygote mice were generated by mating females of the laboratory strain CD1 2n = 40 and males of the Milano II race (2n = 24) or the reciprocal crossings. Mice were maintained at 22 °C with a light/dark cycle of 12/12 hours and fed ad libitum. Procedures involving the use of the mice were reviewed and approved by the Ethics Committee of the Faculty of Medicine, Universidad de Chile (N° CBA #0441) and by the Ethics Committee of the Chilean National Science Foundation FONDECYT-CONICYT.

2.2. Karyotyping

Mitotic chromosomes for karyotype were obtained from bone marrow cells using the conventional method, which includes 0.075 M KCl as a hypotonic solution and methanol acetic acid 3:1 v/v as the fixation solution. Metaphase plates were stained with Giemsa (Merck, Darmstadt, Germany) or with 4′, 6-diamidino-2-phenylindole (DAPI).

2.3. Spermatocyte Microspreads

Spermatocyte spreads were obtained following the procedure described by Peters et al. [17]. Briefly, testicular cells were suspended in 100 mM sucrose for one minute and then spread onto a slide dipped in 1% paraformaldehyde in distilled water containing 0.15% Triton X-100 then left to dry for two hours in a moist chamber. The slides were subsequently washed with 0.08% Photoflo (Kodak, Rochester, NY, USA), air-dried, and rehydrated in PBS. To identify specific chromosomes, the FISH painting technique was performed and, subsequently, a double IF was performed for the detection of the nucleolar protein fibrillarin and the protein SYCP3, a structural component of the synaptonemal complex (SC).

2.4. In-Situ Hybridization (FISH)

Slides containing germ cells prepared as described above were treated for 5 min with PBS, dehydrated in a series of 70, 80, 90, and 100% ethanol for 2 min each and air-dried at room temperature. DNA painting probes specific for nucleolar chromosomes 12, 15 or 16 (Green XCyting Mouse Chromosome Paint Probes, Metasystem) were added to germ cells, covered with coverslips and denatured together at 75 °C for 2 min. Following denaturation, slides were incubated in a humid chamber at 37 °C for 24 h. After incubation, coverslips were removed, and slides rinsed with 0.4 x SSC (saline sodium citrate buffer) at 72 °C for 2 min; 2 x SSC, 0.05% (v/v) Tween20 (Sigma-Aldrich, St. Louis, MO, USA) at room temperature for 30 s. Finally, cells were rinsed twice in PBS for 5 min each. Heterochromatin was stained with DAPI (4′, 6-diamidino-2-phenylindole) (Calbiochem, San Diego, CA, USA) and coverslips mounted with Vectashield.

2.5. Immunofluorescence (IF)

The slides were incubated for 45 minutes at 37 °C in a moist chamber with the primary antibodies: mouse anti-SYCP3 1:100 (Santa Cruz Biotechnology, ab74569) and rabbit anti-Fibrillarin 1:100 (Abcam, Cambridge, UK, ab5821). Then, the slides were incubated for 30 min at room temperature with the secondary antibodies: FITC-conjugated goat anti-mouse IgG (1:50) (Sigma), or Texas red-conjugated goat anti rabbit IgG (1:200) (Jackson ImmunoResearch). Slides were counterstained with 1 μg/mL DAPI (4′, 6-diamidino-2-phenylindole). Finally, slides were rinsed in PBS and mounted in Vectashield (Vector Laboratories).

2.6. Images Analysis

Photographs were obtained using a Nikon (Tokyo, Japan) Optiphot microscope equipped with epifluorescence optics, using the same parameters of exposition and intensity for each picture. Images were analyzed using Adobe Photoshop CS5.1 software (San José, CA, USA) and the Magic Wand Tool to delineate the area of fibrillarin defined by the intensity of red color. The nucleolar expression was established measuring the pixels (px) of the fibrillarin areas. Then, to collectively visualize the bivalents, fibrillarin and heterochromatin, a fusion of the layers was performed using the Lighten tool.

2.7. Statistical Analysis

A comparison of proportions was made with the Z score test using the ratio difference estimator (δ). The normality of the sample distribution in the variables to be tested was determined using the Kolmogorov Smirnov test. For variables without normal distribution, the comparison of means was made with the nonparametric Kruskal–Wallis test. The confidence interval used for all tests was 95% with a level of significance of 5% ($\alpha < 0.05$). All calculations were performed in the Analyze-it® statistical software used in Microsoft Excel® 2010.

3. Results

3.1. FISH and IF in the Detection of Nucleolar Expression in Specific Bivalents of Mouse Spermatocytes

As example of our findings, we show (Figure 1) three representative images of nuclear microspreads of spermatocytes 2n = 40, 2n = 24 and 2n = 32, in which the SCs and the fibrillarin were detected by immunofluorescence and chromosome 15 by FISH. In nuclear microspreads of spermatocytes 2n = 40, it is possible to detect the SCs of the 19 autosomal bivalents and the partial synapses between the sex chromosomes X and Y. The nucleolar bivalent 15, revealed by the green chromatin, showed a non-associated nucleolus (Figure 1a). In spermatocytes 2n = 24, the SCs of 8 metacentric autosomal bivalents, 3 telocentric autosomal bivalents, in addition to the sexual pair in partial synapses, were observed. The Rb nucleolar bivalent 5.15, revealed by the green chromatin in one of its arms, also presented one associated nucleolus (Figure 1b). Finally, in the 2n = 32 spermatocytes, the SCs of 8 autosomal trivalents, 3 autosomal telocentric bivalents and the sexual pair XY in partial synapses were observed. The Rb nucleolar trivalent 5:5.15:15, revealed by the green chromatin in one of its arms, also presented one associated nucleolus (Figure 1c). The trivalents presented a complete synapse between the long arms of the Rb metacentric and the telocentric chromosomes, while the short arms of the latter ones could be unsynapsed or partially synapsed.

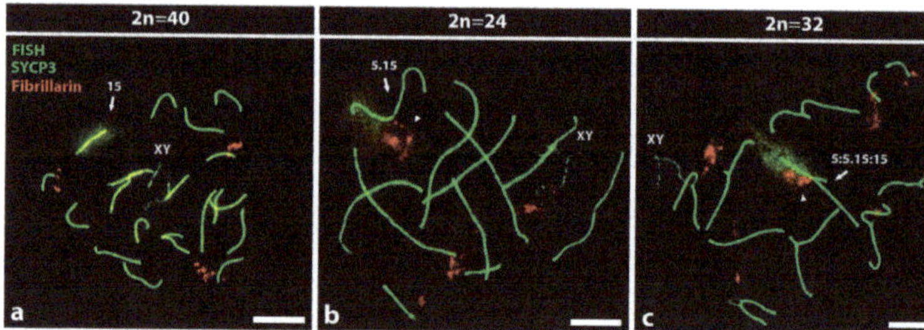

Figure 1. Nuclear microspreads of spermatocytes 2n = 40, 2n = 24 and 2n = 32. By immunocytochemistry, the synaptonemal complex (SCs) of the bivalents (green) and the fibrillarin of the nucleoli (red) are shown. By fluorescent in situ hybridization (FISH) and a specific probe, the chromatin of the nucleolar bivalent 15 (a), or the Rb nucleolar bivalent 5.15 (b) and the trivalent 5:5.15:15 (c), were identified (green). Arrowheads show the nucleoli associated with these nucleolar chromosomes. XY = sex bivalent. Bar = 10 μm. (**a**) The bivalent 15 is indicated among the SCs of nineteen telocentric bivalents, three of them with associated nucleoli. The bivalent 15 does not present associated nucleolus. (**b**) The bivalent 5.15 is indicated among the SCs of 8 metacentric autosomal bivalents, 3 telocentric autosomal bivalents and the sex chromosome pair (XY). The arrowhead indicates the nucleolus associated with the middle of the Rb nucleolar bivalent 5.15. (**c**) The trivalent 5:5.15:15 is indicated among 8 trivalents, 3 telocentric autosomal bivalents and the sex chromosome pair (XY). The arrowhead indicates the nucleolus associated with the middle of the trivalent 5:5.15:15.

To establish the global nucleolar expression between the spermatocytes with different chromosomal constitutions, the pixels (px) of the fibrillarin zones per each nucleus defined by intensity of red color were quantified using the Adobe Photoshop CS5 image software.

The nucleolar expression was higher in 2n = 40 and 2n = 24 spermatocyte nuclei in comparison to 2n = 32 spermatocytes. On average, 4241 px of fibrillarin was estimated in spermatocyte nucleus 2n = 40, 4201 px in 2n = 24 and 3161 px in 2n = 32. The differences between the means of total nucleolar expression of 2n = 40 and 2n = 32 and between 2n = 24 and 2n = 32 were statistically significant ($p < 0.001$). The smallest nucleolar bivalents 18 and 19, present in all the spermatocyte's karyotypes, always expressed large and independent nucleoli, thus constituting a good positive control of nucleolar expression. The FISH method subjects the cells (spermatocytes) to very high temperatures that denature proteins, so their subsequent detection by immunofluorescence is not always possible. To evaluate this negative effect on fibrillarin, 100 nuclear microspreads of spermatocytes 2n = 24, 50 of them treated with FISH and 50 without FISH, were evaluated (not shown). An average of 878 px of nucleolar area associated with nucleolar bivalent 19 was obtained without FISH, whereas an average of 1136 px was obtained with FISH. Between these values, no significant difference was observed (p = 0.0631).

A total of 1800 nuclear microspreads were analyzed, 200 nuclear microspreads for each specific nucleolar chromosome and in the three conditions of spermatocyte's karyotypes. No differences were found between spermatocytes from animals with the same chromosomal condition. The specific nucleolar bivalents were identified by FISH staining and by immunofluorescence of the proteins fibrillarin and SYCP3, respectively, to recognize nucleoli and the SCs of the bivalents (Figure 2). We compared the nucleoli produced by the telocentric nucleolar bivalents 12, 15 and 16 present in 2n = 40 spermatocytes, with those of the respective Rb nucleolar bivalents, Rb 10.12, 5.15 and 16.17 in 2n = 24 spermatocytes, and with those of the respective nucleolar trivalents in 2n = 32 mice (Figure 2). In all cases, multiple nucleoli were observed for each nucleus although the magnitude of the fibrillarin area was variable and with a heterogeneous distribution of intensities. In the telocentric nucleolar bivalent the fibrillarin was observed adjacent to one or both sides of one of the ends of the SC. In the metacentric Rb nucleolar bivalents, fibrillarin was located toward the middle of the SC, whose length was approximately twice that of the telocentric bivalents. In the nucleolar trivalents, the fibrillarin label was located close to the meeting zone of the centromere of the three chromosomes. In spermatocytes 2n = 40, 100% of the nucleolar bivalents 12 had large nucleoli greater than 3 μm in average diameter; only 35% of the bivalents 15 had nucleoli which were small in size, smaller than 1 μm in diameter, and 90% of the nucleolus bivalents 16 had nucleoli of medium size of 2 μm in average diameter (Figure 2a–c). Seventy three percent of the nucleolar bivalents Rb 10.12 and 90% of the nucleolar bivalents Rb 5.15 and 16.17 had medium-sized nucleoli (Figure 2a′–c′). Ninety percent of the trivalents 10:10.12:12 and 5:5.15:15, and 76% of the trivalents 16:16.17:17 had medium nucleoli (Figure 2a″–c″).

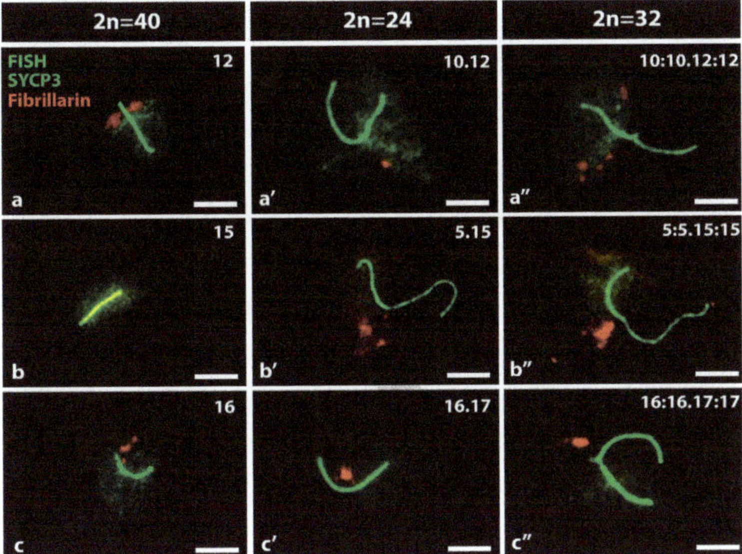

Figure 2. Nucleolar expression in specific ancestral nucleolar bivalents in spermatocytes 2n = 40 and their nucleolar derivatives in spermatocytes 2n = 24 or 2n = 32. (**a**) Bivalent 12; (**a′**) bivalent Rb 10.12; (**a″**) trivalent 10:10.12:12; (**b**) Bivalent 15; (**b′**) bivalent Rb 5.15; (**b″**) trivalent 5:5.15:15; (**c**) Bivalent 16; (**c′**) bivalent 16.17; (**c″**) trivalent 16:16.17:17. By immunocytochemistry, the SCs of the bivalents (green) and the fibrillarin of the nucleoli (red) are shown. By FISH and a specific probe, the chromatin of the nucleolar bivalents, or derivatives, were identified (Green). Bar = 10 μm.

To estimate the nucleolar sizes, the pixels corresponding to the fibrillarin areas of the telocentric bivalents 12, 15 and 16 were quantified and compared with those of the derived chromosomes (Figure 3). The estimated nucleolar expression showed statistically significant differences between bivalent 12 and its derivatives, between bivalent 15 and its derivatives, and between bivalent 16 and only its Rb derivatives. The expression of the NORs of the telocentric bivalents 12, 15 and 16 was variable and much higher in the bivalents 12, although the differences were statistically significant among the three. It is also important to take into account that only 35% of the nucleolus bivalents 15 presented a nucleolus. When comparing the nucleolar expression between the Rb nucleolar bivalents, no significant differences were found, as well as comparing the nucleolar expression between the nucleolar trivalents.

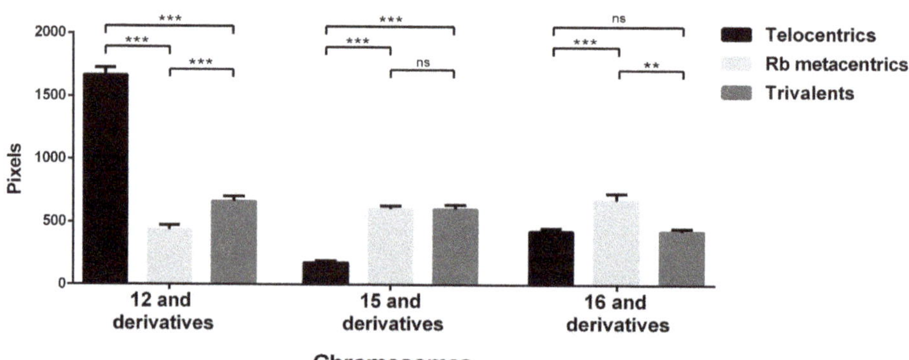

Figure 3. Nucleolar sizes associated with the telocentric bivalents 12, 15 and 16 and their derived chromosomes estimated by the pixels corresponding to the fibrillarin areas. The expression of NORs of bivalents 12, 15 and 16 (black bars) was variable and much higher in the bivalents 12, although the differences were statistically significant among the three. The nucleolar expression between the Rb nucleolar bivalents (light bars) or between the nucleolar trivalents (grey bars) showed no significant differences. The nucleolar expression of bivalent 12 was statistically higher than its derivatives, in bivalent 15 it was statistically lower than its derivatives and in bivalent 16 it was statistically lower than its Rb derivatives and showed no significant differences with the trivalents. Data are expressed as mean $\pm$ SEM: ns: $p > 0.05$, **: $p \leq 0.01$, ***: $p \leq 0.001$ (Kruskal–Wallis test). Rb: Robertsonian.

3.2. Bivalent Associations from Different Shaped Nucleolar Chromosomes Present in Spermatocytes of Mus m. domesticus 2n = 40, 2n = 24 and 2n = 32

The configuration of associations between non-homologous chromosomes in which the specific nucleolar bivalents were involved was studied in spermatocytes 2n = 40, 2n = 24 and 2n = 32. In addition to the specific nucleolar bivalent, the number of SC from other bivalents, the heterochromatin of the associated chromosomes and the presence of other nucleoli were considered (Figure 4). The nucleolar bivalents 12, 15 and 16 had a similar behavior, remaining alone (15%) or in association through heterochromatin with between 2 and 5 bivalents (70%). In one third of these associations there was an additional nucleolus (Figure 4a–c). The Rb nucleolar bivalents were associated through heterochromatin with other metacentric Rb at different approximate frequencies: Rb10.12: 68% associated between 2 and 8 and 32% not associated (Figure 4a′); Rb5.15: 85% associated between 2 and 8 and 15% non-associated (Figure 4b′); Rb16.17: 90% associated between 2 and 8 and 10% non-associated (Figure 4c′). The additional nucleoli were more frequent when increasing the associated bivalents. Nucleolar trivalents were generally isolated (50–70%) or in associations of only two (20–40%) (Figure 4a″–c″). The nucleolar bivalents do not seem to associate preferentially with other nucleolar bivalents and when they participate in the same cluster of heterochromatin the nucleoli are not linked together.

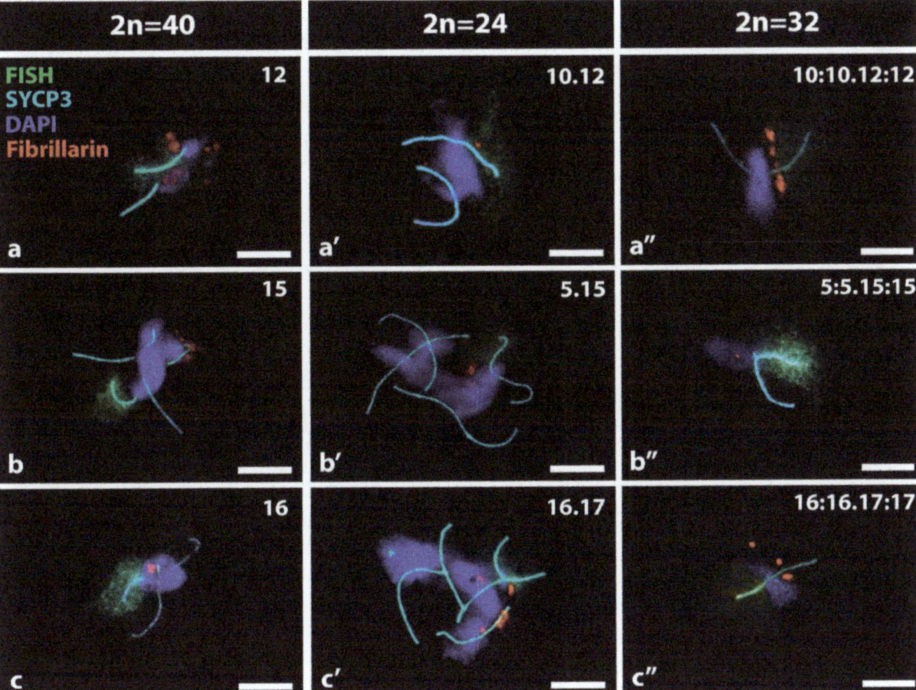

Figure 4. Bivalent associations where a specific nucleolar bivalent is forming part, in spermatocytes 2n = 40 and nucleolar derivatives, in spermatocytes 2n = 24 or 2n = 32. (**a**) bivalent 12; (**a′**) bivalent Rb 10.12; (**a″**) trivalent 10:10.12:12; (**b**) bivalent 15; (**b′**) bivalent Rb 5.15; (**b″**) trivalent 5:5.15:15; (**c**) bivalent 16; (**c′**) bivalent Rb 16.17; (**c″**) trivalent 16:16.17:17. By immunofluorescence the SCs of the bivalents (light blue) and the fibrillarin of the nucleoli (red) are shown. The specific nucleolar bivalents/trivalents were identified by FISH with a probe against the chromatin of the respective nucleolar telocentric chromosome (green). The heterochromatin was stained with DAPI (blue). Bar = 10 μm. (**a**) Bivalent 12 is producing an abundant nucleolus and it is associated through heterochromatin to another bivalent; (**b**) bivalent 15 is not producing a nucleolus and it is associated through heterochromatin to three other bivalents one of them a nucleolar one that has a nucleolus associated; (**c**) bivalent 16 is producing a nucleolus and is associated through heterochromatin to three other bivalents; (**a′**) bivalent 10.12 is producing a small nucleolus and it is associated through heterochromatin to another metacentric bivalent; (**b′**) bivalent 5.15 is producing a medium nucleolus and it is associated through heterochromatin to three metacentric bivalents; (**c′**) bivalent 16.17 is not producing a nucleolus and it is associated through heterochromatin to three metacentric bivalents; (**a″**); (**b″**) and (**c″**) all show one nucleolar trivalent with a nucleolus and not associated with other bivalents at all.

Taking into account that chromosomes of very different lengths are involved in these associations in which at least one nucleolar bivalent was present, we quantified the number of chromosomal arms involved and the results were expressed in a bar graphic (Figure 5). It was observed that the maximum number of chromosomal arms was concentrated in the associations involving Rb nucleolar bivalents and the minimum was concentrated in the territories defined by a nucleolar trivalent (Figure 5). The association of 4 or 5 Rb metacentric bivalents equals the association of 8 or 10 ancestral telocentric bivalents, a value that is much higher than the average of 3 in each cluster of associated bivalents in spermatocyte 2n = 40. Nucleolar trivalents remained isolated in a manner similar to that observed in non-nucleolar trivalents. Each one of the non-associated trivalents compromised a number of chromosomal arms without significant differences in the respective ancestral telocentric bivalents (Figure 5).

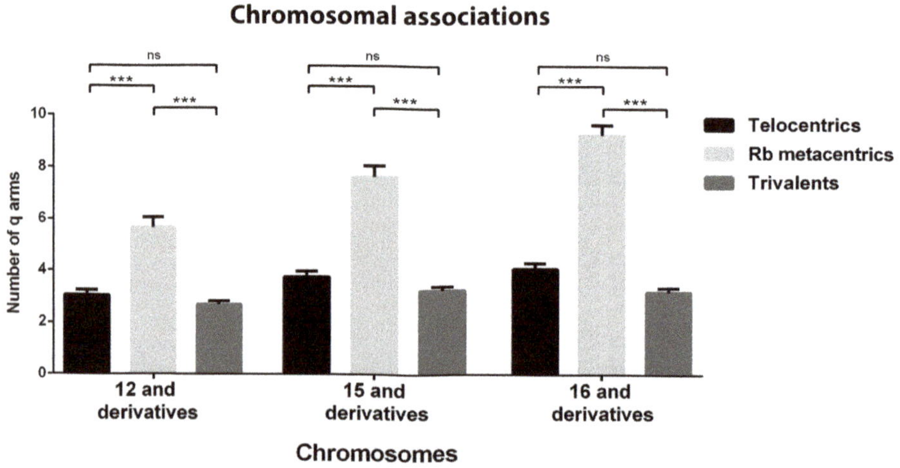

Chromosomal associations

Figure 5. Number of chromosomal arms committed in associations in which a specific nucleolar bivalent is forming part, in spermatocytes 2n = 40 and nucleolar derivatives in spermatocytes 2n = 24 or 2n = 32. In 2n = 40 spermatocytes: an average of 4 chromosome arms were present in chromosomal associations in which NORs of bivalents 12, 15 and 16 were involved (black bars), without significant differences among the three situations. In 2n = 24 spermatocytes: between 6 and 10 chromosome arms were present in chromosomal associations in which NORs of Rb nucleolar bivalents were present (light bars). There were significant differences among the three groups of chromosomal associations. In 2n = 32 spermatocytes: an average of 3 chromosome arms were present in chromosomal associations in which NORs of trivalents were involved (grey bars), without significant differences among the three groups of chromosomal associations. The number of chromosome arms in chromosomal associations defined by a Rb nucleolar bivalent is about twice that observed in the chromosomal associations defined by a telocentric nucleolar bivalent or a nucleolar trivalent. Data are expressed as mean $\pm$ SEM ns = $p > 0.05$, *** = $p \leq 0.001$ (Kruskal–Wallis test).

Through pericentromeric heterochromatin, bivalents of the same morphology associate, whether telocentric or metacentric. In all of these associations, one or more nucleoli surrounded by heterochromatin were observed. Depending on whether the chromosome associations are constituted by telocentric or metacentric bivalents, it would be expected they were distributed respectively in the periphery or in the center of the nuclear space.

4. Discussion

Metacentric Rb chromosomes can become numerous in the *Mus* genome leading to a reduction of ancestral telocentric chromosomes and to an emergence of new mixed karyotypes [11,12,18]. Furthermore, crossing between wild homozygotes 2n = 40 and Rb homozygotes produce F1 hybrids in whose genomes the ancestral telocentric chromosomes are reunited with the metacentric derivatives [19]. During the meiotic prophase I, trivalents with a Rb metacentric chromosome synapsed with two telocentric chromosomes are formed [20,21]. In this work we try to elucidate how or how much gene expression is affected in these new chromosomal and nuclear conditions approaching it through the analysis of NOR expression that produces nucleoli in spermatocytes with different chromosome constitutions. In somatic cells, it is becoming increasingly evident that chromatin organization within the three-dimensional nuclear space (in other words the genome architecture) is itself a likely factor affecting gene regulation and the systemic control of expression of multiple gene loci [22–24]. A similar relationship of the structure and function of the nucleolus has been proposed, whose transcriptional changes would be dependent on the changes in the associated chromatin [25,26]. However, little is known about the organization of nucleolar chromatin in meiotic cells and how this can affect their expression.

The simultaneous application of two methods of difficult compatibility, such as FISH and immunofluorescence (IF), was carried out as has been previously described [27]. Both methods allowed us to identify in each analyzed spermatocyte, a specific nucleolar bivalent and its associated active nucleoli through the identification of fibrillarin protein which is a key factor in nucleolar architecture serving essential functions in rRNA maturation [28]. At the same time, the use of IF for the SYCP3 protein constitutive of the synaptonemal complex, and the use of DAPI, allowed us to observe all the bivalents and if they were associated with each other through the pericentromeric heterochromatin (DAPI).

In 2n = 40 mice, the nucleolar organizing regions (NOR), where the ribosomal genes concentrate, are located in the sub-centromeric region of the long arms of 5 pairs of telocentric chromosomes [4,5]. Therefore, when the Rb fusions involve nucleolar chromosomes, the NORs are structurally preserved [14] unlike the Rb human chromosomes that usually lose the ribosomal genes [29]. Consequently, in the three Rb nucleolar chromosomes of 2n = 24 mice, NORs are located close to the centromeric region and surrounded by the pericentromeric heterochromatin coming from the two original ancestral telocentric chromosomes [14,15]. However, little is known on whether the change of the NOR chromosomal position, and therefore its interaction with other bivalents, affects the magnitude of ribosomal expression in the meiotic prophase. Mice with the 2n = 40 all telocentric karyotype and those carriers of Rb metacentric derived chromosomes are an unbeatable material in the approach to this comparative analysis of nucleolar expression. We observed nucleoli in all the chromosomal conditions studied, and that the observed variability in nucleolar expression of NORs located in the ancestral nucleolar bivalents 12, 15 and 16, was lost in the derived chromosomes. Quantitatively, the nucleolar expression was decreased in the NORs located in the chromosomes derived from 12, increased in the derivatives of 15, and slightly higher in the chromosomes derived from the nucleolar bivalent 16. In the respective derived Rb nucleolar bivalents, as well as in the nucleolar trivalents, the expression of NORs becomes relatively flat. This expression profile could be a consequence of the new environment where NORs are immersed in a greater amount of pericentromeric heterochromatin and in a different disposition with respect to the transcriptional machinery in the nuclear space [10,30]. In most eukaryotes, NORs have an evolutionary conserved positioning in the chromosomes, generally surrounded by constitutive heterochromatin. It has been proposed that such heterochromatin, instead of silencing NORs transcriptionally, may regulate important unknown features of nucleolus formation [31,32]. These, or other suggested mechanisms, that regulate transcription of ribosomal genes, could be affected by the new location of NORs, considering the higher amount of heterochromatin and the apparent different qualities of them [32].

During prophase I of meiosis the SC' organization and trajectory determine the chromosomal domain topology within the nuclear space [33–36]. Considering additionally that the nucleolus persists, bound to the NOR that it originates from [37], interactions or associations among heterologous NORs will depend on the real possibility of establishing contacts between them, particularly in the species with multiple nucleolar chromosomes where these options could be present [10].

Thus, in 2n = 40 mice, the NORs of 5 nucleolar bivalents may interact in the peripheral nuclear space, while in 2n = 24 mice, the NORs of the two nucleolar telocentric bivalents can only interact at the nuclear periphery and the NORs of the three Rb nucleolar bivalents toward the nuclear center. As we have shown here in the spermatocytes 2n = 24, bivalents from these two groups do not interact with each other. With the increase of the Rb fusions in meiotic cells, the real possibilities of interactions or free associations between heterologous chromosomal domains are progressively restricting what can favor or channel the chromosomal or genomic evolution of that karyotype towards a certain path.

We define a nuclear territory in the spermatocytes as a place of the cell nucleus in which topological and functional relationships (or chromosomal associations) are established between bivalents or non-homologous chromosomal domains, which generally originate in the early prophase, frequently mediated by constitutive heterochromatin and can be relatively stable throughout meiotic prophase I [10]. In this work, we selected those chromosomal associations in which specific nucleolar bivalents were involved in order to evaluate whether their morphology or nucleolar expression were gravitating factors in the quality or number of the elements that made up the whole of that chromosomal association.

We did not find a relationship between the magnitude of the nucleolar expression and the number of associated bivalents. Neither was there a preferential aggregation between NORs of different nucleolar bivalents as has been proposed for nucleolar chromosomes considering their structural and functional affinities that could favor the consolidation of such interactions [38,39].

The nucleolar associations were clearly mediated by constitutive pericentromeric heterochromatin and between bivalents of the same morphology, not preferentially between nucleolar bivalents. In other words, the chromosomal associations were conformed by telocentric bivalents or by metacentric bivalents. The associations of 4 or 5 Rb metacentric bivalents equal the association of 8 or 10 ancestral telocentric bivalents, a value that at that same describes the subtracted telocentric chromosomes from the possibilities of associations among them. Focusing on the perspective of the associable bivalents or those that still have possibilities of interaction with others in the nuclear periphery, this is a remarkable change of the original nuclear architecture and that would explain how the new chromosome path is oriented towards the metacentry [40,41].

Trivalents were generally observed not associated with each other or with other bivalents. They and their isolated behavior are an example of the occupation of the same space, in this case 8 trivalents available at the nuclear periphery does not necessarily lead to an association between them. Neither does the presence of abundant heterochromatin by itself or that of active NORs because at least three of the 8 trivalents share all these characteristics and still persist isolated. There is still a lot to understand with respect to what conditions finally make a chromosomal association become consolidated.

It seems clear that chromosomal changes that have already experienced full establishment in a population or reproductive community at the cellular level mean that a huge transformation of the chromosomal associations where they were forming part, and consequently it would be expected they also change the original nuclear architecture. In this sense, Rb nucleolar chromosomes of homozygous mice possibly have been successful since they are multiple and have reached new chromosomal interactions constituting different chromosomal associations, which include new patterns for nucleolar expression.

Author Contributions: Conceptualization, S.B.; methodology, F.L.-M. and E.A.; software, N.Z.; validation, F.L.-M., D.T. and N.Z.; formal analysis, J.L.-F.; resources, S.B.; data curation, E.A. and J.L.-F.; writing—original draft preparation, S.B.; writing—review and editing, S.B. and C.R.

Funding: This work was supported by grants from the Fondo Nacional de Ciencia y Tecnología (FONDECYT; 1190195 (S.B.)) and scholarships from the Comisión Nacional de Investigación Científica y Tecnológica (CONICYT; 21160886 (F.L.-M.)).

Acknowledgments: The authors wish to thank Mr. J Oyarce for his technical assistance and management of the animals to Ann Loughlin for her thorough review of the manuscript.

Conflicts of Interest: The authors declare no conflict of interest.

References

1. Fakan, S.; Hernandez-Verdun, D. The nucleolus and the nucleolar organizer regions. *Biol. Cell* **1986**, *56*, 189–205. [CrossRef] [PubMed]
2. Scheer, U.; Hock, R. Structure and function of the nucleolus. *Curr. Opin. Cell Biol.* **1999**, *11*, 385–390. [CrossRef]
3. Pederson, T. The nucleolus. *Cold Spring Harb Perspect. Biol.* **2011**, *13*, a000638. [CrossRef]
4. Cazaux, B.; Catalan, J.; Veyrunes, F.; Douzery, E.; Britton-Davidian, J. Are ribosomal DNA clusters rearrangement hotspots? A case study in the genus Mus (Rodentia, Muridae). *BMC Evol. Biol.* **2011**, *11*, 124–138. [CrossRef] [PubMed]
5. Britton-Davidian, J.; Cazaux, B.; Catalan, J. Chromosomal dynamics of nucleolar organizer regions (NORs) in the house mouse: micro-evolutionary insights. *Heredity (Edinb).* **2012**, *108*, 68–74. [CrossRef]
6. Berríos, S.; Fernández-Donoso, R.; Pincheira, J.; Page, J.; Manterola, M.; Cerda, M.C. Number and nuclear localisation of nucleoli in mammalian spermatocytes. *Genetica* **2004**, *121*, 219–228. [CrossRef]
7. López-Fenner, J.; Berríos, S.; Manieu, C.; Fernández-Donoso, R.; Page, J. Bivalent associations in *Mus domesticus* 2n = 40 spermatocytes. Are they random? *Bull. Math. Biol.* **2014**, *76*, 1941–1952. [CrossRef]
8. Garagna, S.; Redi, C.; Capanna, E.; Andayani, N.; Alfano, R.; Doi, P.; Viale, G. Genome distribution, chromosomal allocation, and organization of the major and minor satellite DNAs in 11 species and subspecies of the genus Mus. *Cytogenet. Cell Genet.* **1993**, *64*, 247–255. [CrossRef]
9. Guenatri, M.; Bailly, D.; Maison, C.; Almouzni, G. Mouse centric and pericentric satellite repeats form distinct functional heterochromatin. *J. Cell Biol.* **2004**, *166*, 493–505. [CrossRef]
10. Berrios, S. Nuclear architecture of mouse spermatocytes: Chromosome topology, heterochromatin, and nucleolus. *Cytogenet. Genome Res.* **2017**, *151*, 61–71. [CrossRef]
11. Gropp, A.; Winking, H.; Redi, C.; Capanna, E.; Britton-Davidian, J.; Noak, G. Robertsonian karyotype variation in wild house mice from Rhaeto-Lombardia. *Cytogenet. Cell Genet.* **1982**, *34*, 67–77. [CrossRef] [PubMed]
12. Pialek, J.; Hauffe, H.; Searle, J. Chromosomal variation in the house mouse. *Biol. J. Linn. Soc.* **2005**, *84*, 535–563. [CrossRef]
13. Elsevier, S.; Ruddle, F. Location of genes coding for 18S and 28S ribosomal RNA within the genome of *Mus musculus*. *Chromosoma* **1975**, *52*, 219–228. [CrossRef] [PubMed]
14. Garagna, S.; Marziliano, N.; Zuccotti, M.; Searle, J.; Capanna, E.; Redi, C. Pericentromeric organization at the fusion point of mouse Robertsonian translocation chromosomes. *Proc. Natl. Acad. Sci. USA* **2001**, *98*, 171–175. [CrossRef] [PubMed]
15. Berríos, S.; Manieu, C.; López-Fenner, J.; Ayarza, E.; Page, J.; González, M.; Manterola, M.; Fernández-Donoso, R. Robertsonian chromosomes and the nuclear architecture of mouse meiotic prophase spermatocytes. *Biol. Res.* **2014**, *47*, 16–29. [CrossRef] [PubMed]
16. Adriaens, C.; Serebryannyy, L.A.; Feric, M.; Schibler, A.; Meaburn, K.J.; Kubben, N.; Trzaskoma, P.; Shachar, S.; Vidak, S.; Finn, E.H.; et al. Blank spots on the map: some current questions on nuclear organization and genome architecture. *Histochem. Cell Biol.* **2018**, *150*, 579–592. [CrossRef]
17. Peters, A.; Plug, A.; van Vugt, M.; de Boer, P. A drying-down technique for the spreading of mammalian meiocytes from the male and female germline. *Chromosom. Res.* **1997**, *5*, 66–68. [CrossRef]
18. Capanna, E.; Castiglia, R. Chromosomes and speciation in *Mus musculus domesticus*. *Cytogenet. Genome Res.* **2004**, *105*, 375–384. [CrossRef]
19. Garagna, S.; Page, J.; Fernández-Donoso, R.; Zuccotti, M.; Searle, J. The Robertsonian phenomenon in the house mouse: mutation, meiosis and speciation. *Chromosoma* **2014**, *123*, 529–544. [CrossRef]
20. Grao, P.; Coll, M.; Ponsà, M.; Egozcue, J. Trivalent behavior during prophase I in male mice heterozygous for three Robertsonian translocations: an electron-microscopic study. *Cytogenet. Cell Genet.* **1989**, *152*, 105–110. [CrossRef]
21. Manterola, M.; Page, J.; Vasco, C.; Berríos, S.; Parra, M.; Viera, A.; Rufas, J.; Zuccotti, M.; Garagna, S.; Fernández-Donoso, R. A high incidence of meiotic silencing of asynapsed chromatin is not associated with substantial pachytene loss in heterozygous male mice carrying multiple simple robertsonian translocations. *PLoS Genet.* **2009**, *5*, e1000625. [CrossRef] [PubMed]

22. Cremer, T.; Cremer, C. Chromosome territories, nuclear architecture and gene regulation in mammalian cells. *Nat. Rev. Genet.* **2001**, *2*, 292–301. [CrossRef] [PubMed]

23. Misteli, T. Higher-order genome organization in human disease. *Cold Spring Harb. Perspect. Biol.* **2010**, *2*, a000794. [CrossRef] [PubMed]

24. Sexton, T.; Cavalli, G. The role of chromosome domains in shaping the functional genome. *Cell* **2015**, *160*, 1049–1059. [CrossRef] [PubMed]

25. Németh, A.; Grummt, I. Dynamic regulation of nucleolar architecture. *Curr. Opin. Cell Biol.* **2018**, *52*, 105–111. [CrossRef] [PubMed]

26. Lam, Y.; Trinkle-Mulcahy, L.; Lamond, A. The nucleolus. *J. Cell Sci.* **2005**, *118*, 1335–1337. [CrossRef] [PubMed]

27. Scherthan, H. Imaging of chromosome dynamics in mouse testis tissue by immuno-FISH. In *Meiosis*; Humana Press: New York, NY, USA, 2017; pp. 231–243.

28. Farley, K.; Surovtseva, Y.; Merkel, J.; Baserga, S. Determinants of mammalian nucleolar architecture. *Chromosoma* **2015**, *124*, 323–331. [CrossRef]

29. Wolff, D.; Schwartz, S. Characterization of Robertsonian translocations by using fluorescence in situ hybridization. *Am. J. Hum. Genet.* **1992**, *50*, 174–181.

30. Redi, C.; Garagna, S.; Zacharias, H.; Zuccotti, M.; Capanna, E. The other chromatin. *Chromosoma* **2001**, *110*, 136–147. [CrossRef]

31. Peng, J.; Karpen, G. Heterochromatin genome stability requires regulators of histone H3 K9 methylation. *PLoS Genet.* **2009**, *5*, e1000435. [CrossRef]

32. Saksouk, N.; Simboeck, E.; Déjardin, J. Constitutive heterochromatin formation and transcription in mammals. *Epigenetics Chromatin* **2015**, *8*. [CrossRef] [PubMed]

33. Heyting, C. Synaptonemal complexes: structure and function. *Curr. Opin. Cell Biol.* **1996**, *8*, 389–396. [CrossRef]

34. Page, S.; Hawley, R. The genetics and molecular biology of the synaptonemal complex. *Annu. Rev. Cell Dev. Biol.* **2004**, *20*, 525–558. [CrossRef] [PubMed]

35. Fraune, J.; Schramm, S.; Alsheimer, M.; Benavente, R. The mammalian synaptonmemal complex: protein components, assembly and role in meiotic recombination. *Exp. Cell Res.* **2012**, *318*, 1340–1346. [CrossRef]

36. Zickler, D.; Kleckner, N. A few of our favorite things: Pairing, the bouquet, crossover interference and evolution of meiosis. *Semin. Cell Dev. Biol.* **2016**, *54*, 135–148. [CrossRef] [PubMed]

37. Schwarzacher, T.; Mayr, B.; Schweizer, D. Heterochromatin and nucleolus-organizer-region behaviour at male pachytene of *Sus scrofa domestica*. *Chromosoma* **1984**, *91*, 12–19. [CrossRef] [PubMed]

38. Miller, D.; Tantravahi, R.; Dev, V.; Miller, O. Frequency of satellite association of human chromosomes is correlated with amount of Ag-staining of the nucleolus organizer region. *Am. J. Hum. Genet.* **1977**, *29*, 490–502.

39. Pecinka, A.; Schubert, V.; Meister, A.; Kreth, G.; Klatte, M.; Lysak, M.; Fuchs, J.; Schubert, I. Chromosome territory arrangement and homologous pairing in nuclei of *Arabidopsis thaliana* are predominantly random except for NOR-bearing chromosomes. *Chromosoma* **2004**, *113*, 258–269. [CrossRef]

40. Berrios, S.; Manterola, M.; Prieto, Z.; López-Fenner, J.; Fernández-Donoso, R. Model of chromosome associations in *Mus domesticus* spermatocytes. *Biol. Res.* **2010**, *43*, 275–285. [CrossRef]

41. King, M. *Species Evolution: The Role of Chromosome Change*; Cambridge University Press: Cambridge, UK, 1993.

Article

Rescue of *Sly* Expression Is Not Sufficient to Rescue Spermiogenic Phenotype of Mice with Deletions of Y Chromosome Long Arm

Jonathan M. Riel [1], Yasuhiro Yamauchi [1], Victor A. Ruthig [1,†], Qushay U. Malinta [1,‡], Mélina Blanco [2,3,4], Charlotte Moretti [2,3,4,§], Julie Cocquet [2,3,4] and Monika A. Ward [1,*]

[1] Institute for Biogenesis Research, John A. Burns School of Medicine, University of Hawaii, 1960 East-West Rd, Honolulu, HI 96822, USA; jriel@hawaii.edu (J.M.R.); yyamauch@hawaii.edu (Y.Y.); var13@duke.edu (V.A.R.); qushay20019@gmail.com (Q.U.M.)

[2] INSERM, U1016, Institut Cochin, 75013 Paris, France; melina.blanco@inserm.fr (M.B.); charlotte.moretti@ens-lyon.fr (C.M.); julie.cocquet@inserm.fr (J.C.)

[3] CNRS, UMR8104, 75014 Paris, France

[4] Université Paris Descartes, Sorbonne Paris Cité, Faculté de Médecine, 75014 Paris, France

*   Correspondence: mward@hawaii.edu

† Current address: Duke University School of Medicine, Durham, NC 27710, USA.

‡ Current address: Department of Physiology, Faculty of Medicine, Hasanuddin University, Makassar 90245, Indonesia.

Received: 10 January 2019; Accepted: 5 February 2019; Published: 12 February 2019

Abstract: Mice with deletions of the Y-specific (non-PAR) region of the mouse Y chromosome long arm (NPYq) have sperm defects and fertility problems that increase proportionally to deletion size. Mice with abrogated function of NPYq-encoded gene *Sly* (sh367 Sly-KD) display a phenotype similar to that of NPYq deletion mutants but less severe. The milder phenotype can be due to insufficient *Sly* knockdown, involvement of another NPYq gene, or both. To address this question and to further elucidate the role of *Sly* in the infertile phenotype of mice with NPYq deletions, we developed an anti-SLY antibody specifically recognizing SLY1 and SLY2 protein isoforms and used it to characterize SLY expression in NPYq- and *Sly*-deficient mice. We also carried out transgene rescue by adding *Sly1/2* transgenes to mice with NPYq deletions. We demonstrated that SLY1/2 expression in mutant mice decreased proportionally to deletion size, with ~12% of SLY1/2 retained in shSLY sh367 testes. The addition of *Sly1/2* transgenes to mice with NPYq deletions rescued SLY1/2 expression but did not ameliorate fertility and testicular/spermiogenic defects. Together, the data suggest that *Sly* deficiency is not the sole underlying cause of the infertile phenotype of mice with NPYq deletions and imply the involvement of another NPYq gene.

Keywords: Y chromosome; testis; spermatogenesis; SLY; mouse; infertility; assisted reproduction

1. Introduction

Deletions of the Y chromosome are frequently associated with spermatogenic defects both in mice and in humans. In mice, the male-specific, non-pairing Y chromosome long arm (NPYq), (also called MSYq for the male-specific region on the Y chromosome long arm), encompasses ~90% of the Y-specific DNA content and comprises mostly repetitive sequences including multiple copies of four distinct genes that are expressed in spermatids: *Ssty1*, *Ssty2* (Spermiogenesis specific transcript on the Y 1 and 2), *Sly* (Sycp3 like Y-linked), and *Srsy* (Serine rich, secreted, Y-linked) [1]. These multi-copy genes show a progressive reduction in transcript levels with increasing NPYq deficiency and are candidates for contributing to the sperm defects associated with NPYq deletions [2]. Mice with NPYq deletions have sperm defects and are sub- or infertile, with the severity of the phenotype increasing proportionally to the deletion size [3–8]. We succeeded in obtaining live offspring from the infertile males with NPYq deletions when intracytoplasmic sperm injection (ICSI) was used [8,9]; however, the low efficiency of assisted reproduction suggested that sperm impairment reached beyond their inability to transmit the paternal genome to the oocyte in vivo, and might have involved DNA changes. In support of this notion, we have subsequently shown that sperm from mice with severe NPYq deficiencies have DNA damage and abnormal chromatin packaging [10].

To assess which of the NPYq genes is responsible for the infertile phenotype associated with NPYq deficiency, we produced mice in which the function of NPYq-encoded gene *Sly* has been disrupted by transgenically-delivered short hairpin RNAs [11]. The characterization of these 'shSLY mice' (sh367 or Sly-KD for knocked down) revealed infertility, sperm headshape defects, and impairment in sperm chromatin packaging, as well as increased sperm DNA damage, similar to that noted in mice with severe NPYq deletions, but less severe [11,12]. These studies also revealed the underlying cause of Sly-KD and NPYq-spermiogenic phenotypes: Sly-KD or NPYq deletions were shown to lead to a de-repression of sex chromosome-encoded genes and to changes in sex chromatin structure in spermatids [11,13–15]. Molecular analyses showed that SLY1 protein directly regulates the expression of sex chromosome-encoded spermatid-expressed genes, as well as hundreds of spermatid-expressed autosomal genes, with many SLY1 target genes involved in transcriptional regulation and chromatin remodeling [11,14].

Yet, Sly-KD mice phenotype is milder than that of mice with a 9/10th or complete deletion of NPYq. This could be due to insufficient *Sly* knockdown in Sly-KD, involvement of another NPYq gene in the phenotype of mice with NPYq deficiency, or both. To address this question and to further elucidate the role of *Sly* in the infertile phenotype of mice with NPYq deletions, we undertook a two-pronged approach. First, if sperm abnormalities in NPYq-deficient mice are a consequence of *Sly* deficiency, then there should be a correlation between the extent of *Sly* reduction and the severity of sperm defects. We showed earlier that *Sly* transcript levels correlated well with the phenotype [16]. However, the analysis of SLY protein expression was hampered by the fact that the only available SLY antibody only detects the SLY protein long isoform, SLY1, and not the shorter SLY2. To overcome this problem, we developed a new anti-SLY1/2 antibody and used it to characterize SLY expression in NPYq- and *Sly*-deficient mice. Second, if sperm abnormalities in NPYq-deficient mice are a consequence of *Sly* deficiency, then transgenically bringing *Sly*/SLY expression in NPY deficient mice to normal levels, should rescue their infertile phenotype. To address this, we developed mice transgenic for *Sly* and placed the *Sly* transgene in the context of sub- and infertile NPYq-deficient genotypes.

We demonstrated first that Sly-KD mice retain limited quantities of SLY1 and 2 proteins. Importantly, we also showed that males with NPYq deficiency expressing transgenic SLY1 or SLY1/2 at levels comparable to wild-type males displayed fertility impairment and testicular/spermiogenic defects, suggesting the contribution of another NPYq gene to these phenotypes.

2. Materials and Methods

2.1. Chemicals

Pregnant mares' serum gonadotrophin (eCG) and human chorionic gonadotrophin (hCG) were purchased from Calbiochem (San Diego, CA, USA). All other chemicals were obtained from Sigma Chemical Co. (St Louis, MO, USA) unless otherwise stated.

2.2. Mice

Six-to-twelve week-old B6D2F1 (C57BL/6J × DBA/2) females (NCI, Raleigh, NC, USA) were used as oocyte donors for injections and CD-1 (Charles River, Wilmington, MA, USA) or Swiss Webster (NCI) mice were used as vasectomized males and surrogate/foster females for embryo transfer. The mice were fed ad libitum with a standard diet and maintained in a temperature- and light-controlled room (22 °C, 14 h light/10 h dark), in accordance with the guidelines of the Laboratory Animal Services at the University of Hawaii and guidelines presented in National Research Council's (NCR) "Guide for Care and Use of Laboratory Animals" published by Institute for Laboratory Animal Research (ILAR) of the National Academy of Science, Bethesda, MD, 2011. The protocol for animal handling and treatment procedures was reviewed and approved by the Animal Care and Use Committee at the University of Hawaii (animal protocol number 06-010).

The mice of interest in this study were mice with NPYq and *Sly* deficiencies, described by us before [10,12], and mice with transgenic overexpression of *Sly* that were generated in this study. Mice with NPYq/*Sly* deficiencies were on a C57BL/6 genetic background. The XY^{RIII} males on the C57BL/6 background were used as wild-type controls. For a summary of investigated mice, see Table S1.

2.3. Production of Anti-SLY Antibody

To produce an anti-SLY1/2 antibody, a specific peptide (VKSPAFDKNENISPQ) identified by ClustalW alignment of proteins from the XLR family to which SLY belongs, was used to immunize mice using a standard approach. Polyclonal serum with the highest antibody titer was identified by ELISA and screened for specificity using dot blot. The serum that was SLY-specific and recognized both SLY1 and SLY2 isoforms was used to create hybridoma cell lines and transformed into monoclonal anti-SLY antibody. The antibody was further tested for specificity using dot blot, immunofluorescence with HEK293 cells transfected with *Sly1* and *Sly2* ORFs fused to FLAG tags, and Western blot with protein lysates from the testes from mice with NPYq and *Sly* deficiencies.

2.4. Production of SLY1, SLY2, SLX, and SLXL1 Proteins

The open reading frames of *Sly1*, *Sly2*, *Slx*, and *Slxl1* were cloned in-frame with an N-terminal FLAG tag under the control of the CMV promoter of the pCDNA3.1 vector (Invitrogen, Carlsbad, CA, USA). HEK293T cells were transfected using X-treme Gene9 (Roche, Basel, Switzerland) according to the manufacturer's instructions with a pCDNA3.1-CMV vector. Cells were collected for protein extraction 48 h after transfection and pelleted by centrifugation for 5 min at 1000 g. Cells were lysed using ice-cold lysis buffer (50mM Tris HCl with 150 mM NaCl pH 7.4, 1mM EDTA, 1% Triton X-100 and 1 x Complete, Mini, EDTA-free Protease Inhibitor Cocktail). Lysis solution was incubated with anti-FLAG affinity gel (Sigma A2220) at 4 °C overnight on a rotating platform. After centrifugation at 13,000× g at 4 °C for 5 min, the supernatant was discarded and purified protein was eluted from gel.

2.5. Dot-Blot

SLY1, SLY2, SLX, and fetal bovine serum (FBS) proteins were spot dropped on a nitrocellulose membrane and allowed to dry. The membrane was incubated in blocking buffer to prevent non-specific binding and then with an anti-SLY antibody followed by detection with anti-mouse HRP conjugated antibody (sc-2005; Santa Cruz, Dallas, TX, USA) at 1:5000 in PBST. After antibody binding and washes, the membrane was incubated in chemiluminescent solution (SuperSignal West Pico; Thermo Fisher Scientific, Waltham, MA, USA) and imaged by an ImageQuant LAS 4000 biomolecular imager (GE Healthcare Life Sciences, Chicago, IL, USA).

2.6. Immunofluorescence

Immunofluorescence on surface-spread HEK293T cells was performed as previously described [12]. Primary antibodies rabbit anti-FLAG (Sigma, SAB4301135) and SLY (this paper) were diluted 1/500 or used as hybridoma cell culture supernatant, respectively. Secondary antibodies anti-rabbit AF488 and anti-mouse AF488 (Life Technologies Invitrogen, R37118 and A-11001, respectively) were diluted at 1:500. Pictures were taken with an Olympus BX63 wide field fluorescent microscope (Olympus, Tokyo, Japan).

2.7. Western Blotting

Protein extraction and Western blot analyses were performed as described previously [12]. Briefly, 10 to 15 mg of testis or spermatid fraction protein extracts were run on a 12% sodium dodecyl sulfate (SDS)/polyacrylamide gel. Following transfer and blocking, membranes were incubated overnight with one of the following primary antibodies: anti-SLY1 [17], anti-SLY1/2 hybridoma cell culture supernatant, or anti-ACTB at 1:5000. Incubation with the corresponding secondary antibody coupled to peroxidase (anti-mouse IgG sc-2005 or anti-rabbit IgG sc-2313) and detection by chemiluminescence were carried out as described by the manufacturer (SuperSignal West Pico, Pierce Biotechnology, Rockford, IL, USA).

2.8. Production of Mice Transgenic for Sly

Flag-Sly transgenic mice were produced by pronuclear injection of a linearized construct containing either *Sly1* or *Sly2* reading frame, i.e., *Sly1* and *Sly2* isoforms fused with *Flag* sequence, under the control of the spermatid-specific promoter *SP10* (aka *Acrv1*). Fertilized oocytes were microinjected with the construct, using standard protocols. Transgenic founders carrying the *SP10-Flag-Sly1* or *SP10-Flag-Sly2* construct were identified by PCR (primers are shown in Table S2 [11–13,18,19]). The founders with germline transmission were used to propagate transgenic lines. The males from lines with significant *Sly* expression were used to obtain XX*Sly* transgenic females. These females were used for breeding with subfertile 2/3NPYq- males and as oocyte donors for ICSI with sperm from infertile 9/10NPYq- males. The resulting 2/3NPYq-*Sly* and 9/10NPYq-*Sly* males were characterized in respect to their fertility and spermatogenic phenotype. *Sly*-transgenic mice (no FLAG tag) were produced using the same strategy with the construct that did not have the *Flag* sequence.

2.9. Sperm Analyses

To analyze sperm number, motility and morphology, cauda epididymal sperm was released into HEPES-buffered CZB medium (HEPES-CZB [20]), and incubated for at least 10 min at 37 °C immediately before analysis. Sperm counts using a hemocytometer were the mean of three independent scorings per sample. For the analysis of sperm morphology, epididymal sperm were stained with silver nitrate as previously described [8]. Briefly, the sperm suspension (diluted as necessary with 0.9% NaCl) was smeared on three slides, allowed to dry, fixed in methanol and acetic acid (3:1), and stained with silver nitrate. The slides were coded and 100 sperm heads per slide were viewed at 1000× magnification and scored in blind fashion. Categorization of sperm head morphology was performed as previously described [8].

2.10. Assisted Reproduction

Female mice were induced to superovulate with the injection of 5 IU eCG and 5 IU hCG given 48 h apart. Oocyte collection and subsequent oocyte manipulation, including microinjections, were done in HEPES-CZB, with subsequent culture in CZB medium [21] in an atmosphere of 5% CO_2 in air.

To obtain epididymal sperm, caudae epididymides were dissected and sperm were expressed with needles into HEPES-CZB or PBS or T6. Spermatozoa were allowed to disperse for 2–3 min at room temperature. The samples of epididymal cell suspension were used for analyses, for in vitro fertilization (IVF), or for ICSI.

For IVF, the epididymal sperm were capacitated in T6 medium [22] for 1.5 h at 37 °C in a humidified atmosphere of 5% CO_2. In vitro fertilization was performed as previously described [23]. Gametes were co-incubated for 4 h. After co-incubation, the oocytes were washed several times with HEPES-CZB, followed by at least one wash with CZB. Embryos were cultured in CZB and observed at different time points for proper development: 24 h (2-cell stage), 48 h (4-or 8-cell stage), 72 h (morula or early blastocyst), and 96 h (blastocyst).

Intracytoplasmic sperm injection was carried out as previously described [24], within 1–2 h from oocyte and sperm collection. Sperm injected oocytes were transferred in CZB and cultured at 37 °C. The survival and activation of injected oocytes was scored 1–2 h and 6 h after the commencement of culture, respectively. The oocytes with two well-developed pronuclei and extruded second polar body were considered activated.

Embryos resulting from ICSI that reached the 2-cell stage were transferred to the oviducts (10–14 per oviduct) of CD1 females mated during the previous night with vasectomized CD1 males. Surrogate mothers were allowed to deliver and raise their offspring or had a cesarean section performed and the progeny raised by foster mothers. The progeny were genotyped after weaning (age, 21 days) and subsequently used for the analyses of fertility and spermiogenic phenotype.

2.11. Real-Time RT-PCR

For real-time reverse transcriptase polymerase chain reaction (RT-PCR), total testis RNA was extracted using Trizol and DNaseI treatment (Ambion, Austin, TX, USA), and purified using an RNeasy kit (Qiagen, Valencia, CA, USA). Reverse transcription of polyadenylated RNA was performed with Superscript Reverse Transcriptase IV, according to the manufacturer's guidelines (Invitrogen). Real-time PCR was performed using SYBR Green PCR Master mix on an ABI QuantStudio 12K Flex machine (Applied Biosystems, Carlsbad, CA, USA). PCR reactions were incubated at 95 °C for 10 min followed by 35 PCR cycles (10 s at 95 °C and 60 s at 60 °C). For analysis of *Sly* expression, two types of PCR reactions were performed: (1) 'Sly1+2' amplifying both *Sly*_v1 and *Sly*_v2 transcripts [17] (primers *Sly* Global) and (2) 'Sly1' amplifying only *Sly*_v1 (primers *Sly* Long). All reactions were carried out in triplicate per assay and *Actb* was included as a loading control. The ΔCt value for each individual sample was calculated by subtracting the average ΔCt of *Actb* from the average ΔCt of each tested gene. ΔΔCt was calculated by subtracting the ΔCt of each tested male from the average ΔCt of the reference samples (non-transgenic siblings). The data were expressed as a fold value of expression level. Expression analysis was also done on several X and Y encoded transcripts to test for their upregulation in NPYq-deficient mice with and without the *Sly* transgene. Primer sequences are shown in supplementary material Table S2.

2.12. Development and Analyses of SP4 Transgenic Mice

An additional transgenic line expressing *Sly1* (*Flag-Sly1* SP4, subsequently called SP4) was developed and characterized independently in Cocquet lab in France. Overall, a similar approach and methodology were used as those described above for mice generated in Ward lab in Hawaii. The few differences are as follows:

Mice. All animals were on ~90% C57BL/6 background and processed at adult age (between 2- and 6-month-old males). The SP4 line was obtained by pronuclear micro-injection of a linearized construct containing a *Sly1* open reading frame fused with a *Flag* sequence, under the control of the spermatid-specific promoter *SP10* (aka *Acrv1*) [14]. Transgenic SP4 *Flag-Sly1* males with a wild-type Y chromosome (i.e., XY^{RIII} or XY^{B10}) have already been described [14]. In brief, RT-qPCR and Western blot experiments showed a ~2-fold increase in *Sly1* transcript and SLY1 protein level in SP4 transgenic testes compared to testes from non-transgenic (WT) siblings. The 2/3NPYq- (2/3MSYq-) mice have a Y^{RIII} chromosome with a deletion removing approximately two-thirds of the NPYq. NPYq- (MSYq-) mice are XSxr^{a}Y*^{X} mice [3] and lack the entire Y-specific (non-PAR) gene content of NPYq, with the only Y-specific material provided by the Y short-arm-derived factor *Sxr^a*. To produce SP4 transgenic males with a 2/3NPYq-, SP4 transgenic females were mated to 2/3NPYq- males. To produce SP4 transgenic males with NPYq deletion, SP4 transgenic females were mated to XYSxr^a males. Then SP4 transgenic XYSxr^a males were mated to XY*^{X} females to produce SP4 transgenic XSxr^{a}Y*^{X} males (i.e., SP4 transgenic NPYq- males) [25–28]. The animal procedures were in accordance with the United Kingdom Animal Scientific Procedures Act 1986 and subject to local ethical review in UK and France

(Comite d'Ethique pour l'Experimentation Animale, Universite Paris Descartes; registration number CEEA34.JC.114.12).

Real-time quantitative PCR. For the quantification of *Flag-Sly* transgene expression, total RNA was extracted from adult testes and reverse-transcribed as described above. Real-time PCR was performed using Roche Light Cycler 480 and Bioline SensiFAST SYBR No-ROX Kit (Meridian Bioscience, Paris, France) using the primers described in Table S2. *Acrv1* was included on every plate as a loading control. *Acrv1* expression was not affected in mice with Y chromosome deletion or Sly-KD mice, checked by both microarray and qRT-PCR [11].

Western blot and immunofluorescence. Western blot and immunofluorescence analyses were performed using the same conditions as described before [29]. For Western blot analyses, anti-SLY1 antibody [17] was used at 1/3000; for immunofluorescence experiments, anti-SLY1 was used at 1/100. Alexa Fluor 594-conjugated peanut agglutinin lectin (Invitrogen) stains the developing acrosome and was used at 1/500 to stage testis tubules.

Analysis of sperm head morphology. Silver staining of sperm smears obtained from the initial caput epididymis was performed as described previously [7]. The analyses were performed in blind fashion. In comparison with classification of the Ward lab [8], the categories were as follows: slightly flattened = 1S + 2S, grossly flattened = 4G, and other gross abnormalities = 3G + 5G + 6G + 7G + 8G.

2.13. Statistics

The transcript and protein expression, testis and body weight, and sperm number were analyzed with a *t*-test. Fertilization rate data were analyzed by Fisher's Exact Test. Sperm headshape defects data were assessed by two-way ANOVA using the GraphPad Prism version 5.0 software after transforming all percentages to angles.

3. Results

3.1. New Anti-SLY Antibody Specifically Recognizes SLY1 and SLY2 Proteins

Thus far, analyses of SLY protein expression have been hampered by a lack of suitable antibody because the previously used serum recognized only one of the two existing SLY isoforms, SLY1. To overcome this limitation, we raised an antibody that specifically recognized both SLY1 and SLY2.

A ClustalW alignment of XLR, SLX, SLX1, and SYCP3 with the predicted SLY1 and SLY2 amino acid sequences (Figure S1) identified a 15-amino acid, putatively SLY1- and SLY2-specific peptide. This peptide was used to generate a monoclonal anti-SLY1/2 antibody. The antibody was able to detect SLY1 and SLY2 proteins in HEK293 cells transfected with *Sly1* and *Sly2* ORF fused to the *Flag* tag but did not react with HEK293 cells transfected with *Flag-Slx* and *Slxl1* ORF (Figure S2). The antibody also detected SLY1 and SLY2, but not SLX, on dot blots (Figure S3).

To further verify that the antigen is the putative SLY1 and SLY2 protein, Western blot analyses were carried out on testis extracts from males with previously reported [11,12] deficiencies in *Sly* transcript expression (Figure 1). Based on prior quantification of the *Sly* transcript level, *Sly*-deficient transgenic line sh344 was expected to have *Sly1*-specific deficiency while transgenic line sh367 was expected to be deficient for both *Sly1* and *Sly2*, and be more strongly affected overall [12]. The Western analyses were in agreement with the transcript expression data; sh344 males showed severe deficiency of SLY1 but not SLY2 while sh367 males had no SLY1 and severe loss SLY2 (Figure 1A,B).

Combined results testing the specificity of a newly developed antibody supported that it specifically recognizes SLY1 and SLY2 and is therefore appropriate to comprehensively evaluate SLY expression.

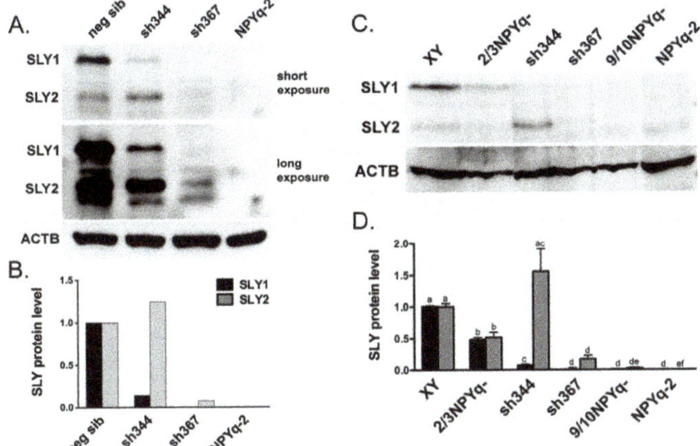

Figure 1. SLY expression in males with NPYq- and *Sly*-specific deficiency. (**A,B**) New anti-SLY antibody specifically recognizes SLY1 and SLY2 isoforms. (**A**) Exemplary Western blot detection of SLY1 (~40 kDa) and SLY2 (~30 kDa) protein in testis lysates from Sly-KD transgenic mice with *Sly* deficiency (sh344 and sh367). The positive control was a negative sibling of Sly-KD mice (neg sib) while the negative control was a male lacking the NPYq (NPYq-2). (**B**): Levels of protein expression shown in panel A quantified with *ImageJ* software (https://imagej.nih.gov/ij/) and normalized with respect to ACTB signal and with neg sib data serving as the normal expression baseline. (**C,D**) SLY expression in males with NPYq- and *Sly*-specific deficiency. (**C**) Exemplary Western blot detection of SLY1 and SLY2 protein in testis lysates from wild-type control (XY), mutant mice with progressively increasing NPYq deficiency (2/3NPYq-, 9/10NPYq-, and NPYq-2) and sh344 and sh367. (**D**) Levels of protein expression quantified with *ImageJ* software and normalized with respect to the ACTB signal and with XY data serving as the normal expression baseline. The data represent average ± SEM of several western runs with the following male number per genotype: n = 8, 6, 6, 6, 4, 7 for XY, 2/3NPYq-, sh344, sh367, 9/10NPYq-, and NPYq-2, respectively. Statistical significance (*t*-test): for each protein isoform, the genotypes marked with different letters are different from each other.

3.2. SLY Protein Expression in Mice with NPYq and Sly Deficiencies Matches Previously Reported Transcript Expression Data

The expression of SLY1 and SLY2 in the testes from males with NPYq- and *Sly*-deficiencies was examined using several males per genotype and repeated western runs (Figure 1C,D and Figure S4). Both SLY1 and SLY2 were reduced about 2-fold in the testes from males carrying a partial NPYq deletion (2/3NPYq-). Males with the deletion removing 9/10 of the NPYq (9/10NPYq-) had negligible SLY1 and SLY2 and males lacking the entire NPYq (NPYq-2) lacked both SLY isoforms. The repeated analyses of sh344 and sh367 transgenic males confirmed that the sh344 males overexpressed SLY2 and retained some of SLY1 while the sh367 males had an almost complete loss of SLY1 and very significant reduction of SLY2. Overall, the protein expression data were in agreement with the previously reported [12] decrease in transcript levels in NPYq- and *Sly*-deficient mice.

3.3. Addition of Flag-Sly Transgenes Rescues Sly/SLY Expression Deficiency in 2/3NPYq- and 9/10NPYq-Mice

To further investigate the role of SLY in the spermiogenic phenotype of mice with NPYq/*Sly* deficiencies, we pursued a transgene rescue strategy. First, mice transgenic for two *Sly* transcript variants, *Sly1* and *Sly2*, fused with *Flag* tag sequence and under the control of the spermatid-specific promoter *SP10*, were produced by pronuclear injection. *Sly1*- and *Sly2*-specific constructs were injected either separately (to yield single gene, *Sly1* or *Sly2*, transgenics) or together (to yield double-gene, *Sly1/2*, transgenics). The offspring derived from pronuclear injection were genotyped and two *Sly1/2* double transgenic, six *Sly1* transgenic, and two *Sly2* transgenic founders with germline transmission were obtained (Figure S5A). Male F1 and F2 progeny derived from these founders provided testes for *Sly* expression analyses. These analyses identified one *Sly1/2* (6P) and two *Sly1* (30A and 16D) transgenic lines with *Sly* overexpression 2–8-fold higher when compared to endogenous *Sly* levels (Figure S5B). These three lines were propagated to produce transgenic females.

Transgenic females were used for breeding with subfertile 2/3NPYq- males and provided oocytes for ICSI with sperm from infertile 9/10NPYq- males. The resulting NPYq-deficient male progeny were genotyped and the males with the supplementing *Flag-Sly* transgene/s (tsgic) and their transgene negative siblings (neg sib) were examined for the rescue of *Sly* expression.

The addition of the *Flag-Sly* transgene partially or completely rescued *Sly* expression in all three transgenic lines. When the *Flag-Sly* transgenes were added to 2/3NPYq-deficient mice (Figure 2A and Figure S6), all three transgenic lines had *Sly* levels significantly higher than their transgene negative siblings. When 2/3NPYq- *Flag-Sly* males were compared to wild-type control (XY), there were no differences in *Sly1/2* and *Sly1* levels between XY and transgenic males from line 30A (complete expression rescue) while males from transgenic lines 6P and 16D displayed partial *Sly* rescue (6P: partial *Sly1* and complete *Sly1/2*; 16D: complete *Sly1* and partial *Sly1/2*). When the *Flag-Sly* transgenes were added to 9/10NPYq-deficient mice (Figure 2B and Figure S6), all transgenic males had *Sly* levels significantly higher than their negative siblings. When 9/10NPYq- *Flag-Sly* males were compared to the wild-type control (XY), there were no differences in *Sly1/2* and *Sly1* levels between XY and transgenic males from line 6P (complete expression rescue) while transgenic males from lines 30A and 16D had partial rescue of both *Sly1/2* and *Sly1* expression.

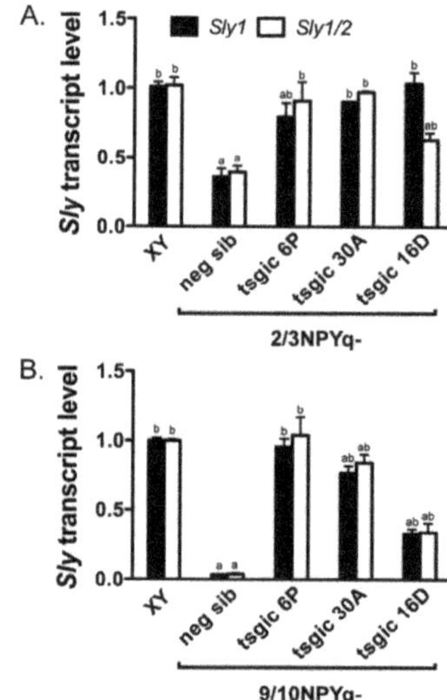

Figure 2. Addition of the *Flag-Sly* transgene to males with NPYq deletions rescues *Sly* expression deficiency. *Sly* transcripts levels (*Sly1* and *Sly1/2* global) in whole testes from moderately (**A**) and severely (**B**) NPYq-deficient mice with (tsgic) and without (neg sib) *Flag-Sly* transgene addition obtained by real-time RT-PCR with *Actb* as a loading control and normalized to wild-type XY controls. Three transgenic lines were tested: 6P carrying *Sly1* and *Sly2* transgenes and lines 30A and 16D positive for *Sly1* transgene only. There were no differences between XY and neg sib from different transgenic lines so the XY and neg sib data were pooled (the data showing all transgenic lines assayed separately is shown in Figure S6). The graphs are mean ± SEM with n = 14, 14, 9, 3, 4 (**A**) and n = 9, 9, 3, 3, 3 (**B**) for XY, neg sib, and tsgic 6P, 30A and 16D, respectively. Statistical significance (t-test, $p < 0.05$): [a] different than respective transcript type in XY; [b] different than respective transcript type in neg sib. Primer sequences are shown in Table S2.

Western blot analyses were performed to confirm that transcript expression rescue translated to the rescue of SLY protein expression. Because the production and analyses of mice with NPYq deficiency and transgene rescue preceded our development of a new anti-SLY1/2 antibody, we first tested SLY1 rescue using an anti-SLY1 antibody developed and reported on before [17]. Western blots were obtained with testis lysates from XY males and 2/3NPYq- males with (tsgic) and without (neg sib) *Flag-Sly* transgene addition (Figure 3 and Figure S7). Two transgenic lines were tested, 16D positive for the *Sly1* transgene and 6P positive for both the *Sly1* and *Sly2* transgenes. 2/3NPYq- males had SLY1 levels 3–4-fold lower than XY, as expected. In 2/3NPYq-*Sly* transgenic males SLY1 expression deficiency was rescued yielding SLY1 levels similar to those of XY controls (Figure 3 and Figure S7).

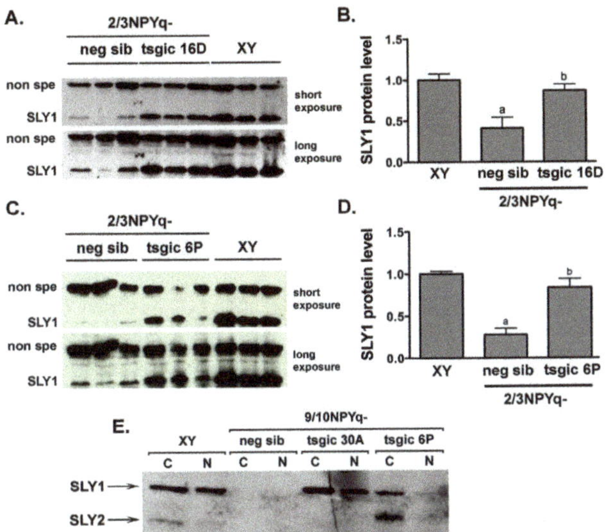

Figure 3. Addition of the *Flag-Sly* transgenes to males with NPYq deficiency rescues SLY1 and SLY2 expression deficiency. (**A–D**) Western blot was performed with whole testes lysates obtained from XY males and from males with moderate NPYq deficiency (2/3NPYq-) with (tsgic) and without (neg sib) *Flag-Sly* transgene addition. Two transgenic lines were tested: 16D positive for the *Sly1* transgene (**A,B**) and 6P positive for both the *Sly1* and *Sly2* transgenes (**C,D**). Levels of protein expression shown in panels (**A,C**) were quantified with *ImageJ* software, normalized to a non-specific band (**B,D**) with XY data serving as the normal expression baseline. The normalization for line 6P was also done to Ponceau signal (Figure S7). The data represent an average ± SEM with $n = 3$. Statistical significance (*t*-test, $p < 0.05$): [a] different from XY; [b] different from neg sib. (**E**) Western blot was performed with cytoplasmic (C) and nuclear (N) protein lysates from the whole testes obtained from XY males and males with severe NPYq deficiency (9/10NPYq-) with (tsgic) and without (neg sib) *Sly* transgene addition. Two transgenic lines were tested: 30A positive for the *Flag-Sly1* transgene and 6P carrying both the *Sly1* and *Sly2* transgenes. Due to the scarcity of testicular material, the reference gene was not included in this analysis.

The rescue of SLY expression was also examined in mice with severe NPYq deficiency (9/10NPYq-) with (tsgic) and without (neg sib) *Flag-Sly* transgene addition, with a distinction between cytoplasmic and nuclear SLY expression to assess the transgenic SLY ability to mimic the behavior of endogenous SLY (Figure 3E). When Western blot was performed with cytoplasmic and nuclear protein lysates from whole testes obtained from wild-type XY males, SLY1 was abundantly present in both cytoplasmic and nuclear fractions, while SLY2 was much less abundant overall (Figure 3E). No SLY1 and SLY2 were detected in mice with severe NPYq deficiency (9/10NPYq-), as expected. However, when the *Flag-Sly* transgenes were added, the SLY1 expression was rescued in 9/10NPYq- males transgenic for

Sly1 (30A), and both SLY1 and SLY2 expression was rescued in 9/10NPYq- males transgenic for *Sly1/2* (6P). Similarly, as in XY, SLY2 predominated in the cytoplasmic fraction. Because of the scarcity of testicular material, only one Western blot was performed and a reference gene was not included in this analysis. However, even with this limitation, the obtained results clearly show that transgenic SLY1 and SLY2 proteins were present and capable of entering into the cell nuclei (Figure 3E).

3.4. Addition of Flag-Sly Transgenes Does Not Rescue Spermiogenic Phenotype of 2/3NPYq- and 9/10NPYq-Mice

2/3NPYq-*Sly* and 9/10NPYq-*Sly* males were subjected to the analyses of rescue of sub- and infertile phenotype via the analysis of the sperm parameters (number, motility and morphology) and sperm ability to fertilize oocytes in vitro. The complete characterization was done for 9/10NPYq-*Sly* males (Table 1, Figure 4) while for 2/3NPYq-*Sly* males, only sperm morphology, the most prominent spermiogenic phenotypic feature of these mice, was assessed (Figure 4).

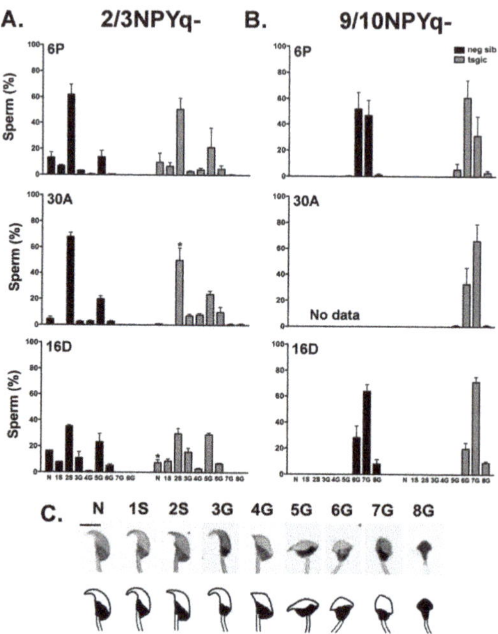

Figure 4. Addition of the *Flag-Sly* transgene does not rescue sperm morphology defects in males with NPYq deficiency. Sperm headshape was evaluated in mice with moderate (**A**, 2/3NPYq-) and severe (**B**, 9/10NPYq-) deficiency with (tsgic) and without (neg sib) the *Flag-Sly* transgene. Three transgenic lines were tested: 6P positive for both the *Sly1* and *Sly2* transgenes (top), and 30A (middle) and 16D (bottom) positive for the *Sly1* transgene. Normal headshape (N) and eight categories of headshape defects (slight: 1S-2S and gross: G3–G8) were differentiated (**C**). The data represent an average ± SDev with n = 2–4 males per genotype and 300 sperm examined per male (**A**) or average ± SDev with n = 3–4 males per genotype and 100 sperm examined per male (**B**). The photo/diagram composite (**C**) was published by us before (Figure 4 in Reference [8]); Bar = 5 μm. No data for neg sib are shown for line 30A because no transgene negative siblings were obtained in ICSI trials. Statistical significance: two-way ANOVA with genotype and sperm headshape as factors, revealed no effect of genotype (p > 0.05) and a strong effect of headshape (p < 0.0001) for all groups tested. The interaction effect (p < 0.05) was present only for 2/3NPYq- 30A and 16D groups. The results of paired comparison for specific sperm headshape types between transgenic and negative siblings in a post-hoc Bonferroni test are shown within the graphs: * p < 0.05.

Table 1. Addition of the *Flag-Sly* transgene to males with severe NPYq deficiency does not rescue low sperm number and sperm ability to fertilize oocytes in vitro.

Tg Line	Genotype	No Males	Age (Weeks)	Average $\pm$ SEM			Eggs Fertilized/Eggs Inseminated (%)
				Body Weight (g)	Testis Weight (mg)	Sperm Number (1CE, $\times 10^6$)	
6P	9/10NPYq-*Sly1/2*	4	10	27 $\pm$ 1	58 $\pm$ 4	0.1 $\pm$ 0.1	0/236 (0)
	9/10NPYq-	3	10	26 $\pm$ 2	60 $\pm$ 3	1.9 $\pm$ 0.9	0/167 (0)
	WT IVF control	2	16	34 $\pm$ 0	92 $\pm$ 3	n/a	75/121 (62)
16D	9/10NPYq-*Sly1*	4	12	32 $\pm$ 4	79 $\pm$ 10	2.5 $\pm$ 1.2	1/236 (0)
	9/10NPYq-	3	12	27 $\pm$ 1	81 $\pm$ 2	4.2 $\pm$ 2.4	0/166 (0)
	WT IVF control	2	14	29 $\pm$ 1	94 $\pm$ 3	n/a	114/122 (93)
30A	9/10NPYq-*Sly1*	3	10	26 $\pm$ 1	64 $\pm$ 4	0.4 $\pm$ 0.03	0/129 (0)
	WT IVF control	1	16	38	99, 100	n/a	39/49 (80)

Tg = transgene. 1CE = 1 cauda epididymis. There were no statistically significant differences between 9/10NPYq-*Sly* transgenics and their negative siblings 9/10NPYq- for any of the factors tested.

There were no differences in body weight, testis weight, and sperm number between 9/10NPYq-*Sly* males and their transgene negative 9/10NPYq- siblings in all three transgenic lines tested (6P, 16D, and 30A). Both 9/10NPYq- and 9/10NPYq-*Sly* males had low sperm number (range: 0.1×10^6–4.2×10^6) typical of the 9/10NPYq- genotype, and these sperm were unable to fertilize oocytes in vitro (Table 1).

Sperm headshape defects are the most prominent feature of mice with NPYq deficiencies. The incidence of various headshape defects in mice with moderate (2/3NPYq-) and severe (9/10NPYq-) NPY deficiency with (tsgic) and without (neg sib) *Flag-Sly* transgenes was quantified according to criteria established by us before [8] (Figure 4). In agreement with previously published data [8,9], 2/3NPYq- mice were moderately affected with ~20% of sperm having a normal headshape and the remaining ~80% having various headshape defects, the majority of which were categorized as slight (Figure 4A), while 9/10NPYq- males had no morphologically normal sperm and all observed headshape defects were gross (Figure 4B). The presence of the *Flag-Sly* transgenes did not rescue sperm headshape abnormalities, with transgenic males presenting with a similar frequency and distribution of defects as their transgene negative siblings (Figure 4).

3.5. Addition of Flag-Sly Transgenes Does Not Rescue Gene Upregulation Associated with 2/3NPYq- and 9/10NPYq-Deficiency

NPYq- and *Sly*-deficient males were shown before to display a remarkable upregulation of sex chromosome genes after meiosis, and we proposed that the spermiogenic defects associated with NPYq/*Sly* deficiency might be a consequence of the massive and global upregulation of spermiogenic genes [11,12]. Since 2/3NPYq-*Sly* and 9/10NPYq-*Sly* males displayed a similar spermiogenic phenotype as their non-transgenic siblings, we predicted that both types of males will also present with the X and Y gene upregulation. The analysis of expression of several X (*Slx*, Sycp3 like X-linked; *Slxl1*, Slx-like 1; *Astx*, Amplified spermatogenic transcript X encoded 5; *Mgclh*, Germ cell-less protein-like 2; *Actrt1*, Actin-related protein T1; *Tcp11x2*, T-complex 11 family, X-linked 2), Y (*Zfy2*, Zinc finger protein 2, Y-linked) and autosomally (*Ubb*, Ubiquitin B; *Tnp1*, Transition protein 1; *Prm1*, Protamine 1) encoded genes revealed that genes were upregulated in both 9/10NPYq- and 9/10NPYq-*Sly* males (Figure 5).

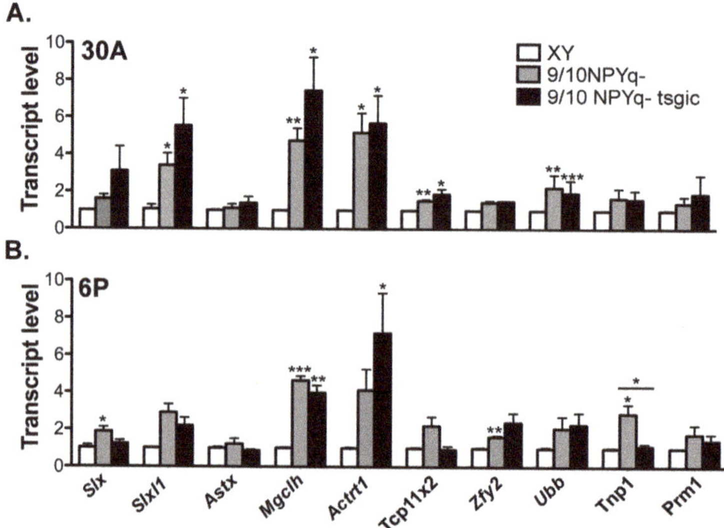

Figure 5. Addition of the *Flag-Sly* transgene to males with NPYq deletions does not rescue X-Y gene upregulation. Gene expression in testes from NPYq-deficient mice without (9/10NPYq-) and with (9/10NPYq- tsgic) the *Flag-Sly* transgene was obtained by real-time RT-PCR with *Actb* as a loading control and normalized to wild-type XY controls. Two transgenic lines were analyzed: line 30A carrying the *Sly1* transgene (**A**) and line 6P carrying both the *Sly1* and *Sly2* transgene (**B**). NPYq- deficient mice without the transgenes were negative siblings of transgenics. The graphs are mean ± SEM with *n* = 3. Statistical significance (*t*-test): * < 0.05; ** < 0.01; *** < 0.001, in comparison with NPYq- deficient mice with XY; the difference between transgenics and their negative siblings is marked with a horizontal lane above the relevant graph pair. Primer sequences are shown in Table S2.

3.6. Flag-Sly Transgenic Line Generated and Analyzed Independently Confirms that Addition of the Sly1 Transgene Does Not Rescue Spermiogenic Phenotype of Mice with NPYq Deficiency

All data presented thus far were obtained with transgenic and NPYq/*Sly* deficient mice produced in Hawaii (Ward Lab). Independently, in France (Cocquet Lab), another line of mice transgenic for *Flag-Sly1*, SP4, were generated and examined. Transgenic XY SP4 males had *Sly1* and global *Sly1/2* transcript levels elevated 2–2.5-fold compared to non-transgenic siblings and the SP4 *Sly1* transgene recapitulated all features of endogenous SLY1 in the XY context [14]. When the SP4 *Flag-Sly1* transgene was added to 2/3NPYq- mice, *Sly1* expression was rescued at the transcript level (Figure 6A). Western blot analyses were performed and showed that the SLY1 protein level is also rescued with a protein level similar in 2/3NPYq- with the SP4 *Sly1* transgene and in WT (XY) testes (Figure 6B). When the SP4 *Flag-Sly1* transgene was added to mice lacking all NPYq genes (NPYq-), the SLY1 protein level of expression was also rescued (Figure 6C) and the pattern of transgenic FLAG-SLY1 expression was similar to that of endogenous SLY1 in XY testes, i.e., strong nuclear and cytoplasmic signal in steps 2/3 to 9 spermatids (Figure 6D) [11,17]. The presence of the SP4 *Flag-Sly1* transgene in 2/3NPYq- males did not rescue the sperm headshape defects (Figure 6E) and derepression of X encoded genes *Slxl1*, *Tcp11x2* and *Mgclh* (Figure 6A). Similarly, SP4 *Flag-Sly1* transgene in NPYq- males did not rescue testis weight nor sperm number (average NPYq- SP4 testis weight = 67.5 mg vs. WT = 102.3 mg; average NPYq- SP4 sperm number per cauda epididymis = 0.45 × 10⁶ vs. WT = 13.6 × 10⁶). Altogether, these data confirm by an independent study that the rescue of SLY1 expression level and pattern with a *Flag-Sly1* transgene does not rescue the NPYq deficiency phenotype.

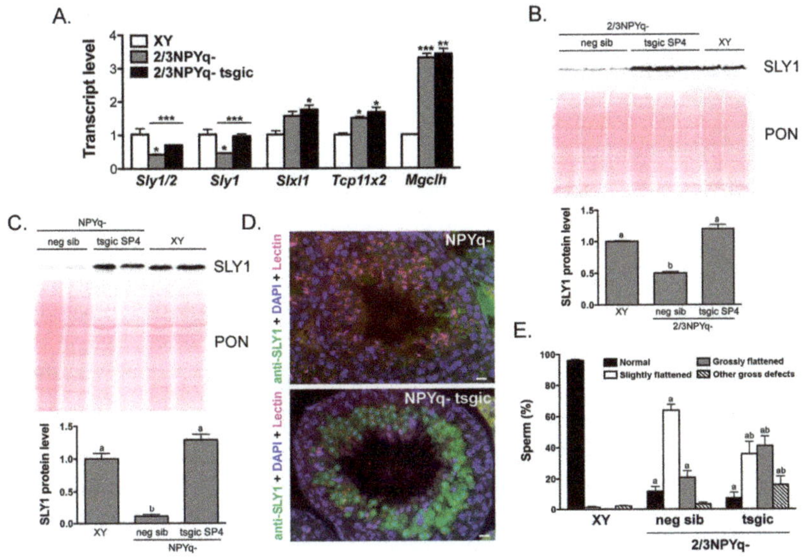

Figure 6. Flag-*Sly1* SP4 transgenic line generated and analyzed independently confirms that the rescue of *Sly1* expression does not lead to the rescue of the spermiogenic phenotype of mice with NPYq deficiency. (**A**) Gene expression in testes from NPYq-deficient mice without (2/3NPYq-) and with (2/3NPYq- tsgic) the *Flag-Sly1* SP4 transgene was obtained by real-time RT-PCR with *Acrv1* as a loading control and normalized to wild-type XY controls. The graphs are mean $\pm$ SEM with $n = 3$. Statistical significance (t-test): * < 0.05; *** < 0.001, in comparison with NPYq-deficient mice with XY; the difference between transgenics and their negative siblings is marked with a horizontal lane above the relevant graph pairs. Primer sequences are shown in Table S2. (**B,C**) SLY1 expression rescue shown with Western blot performed with whole testes lysates obtained from males with moderate NPYq deficiency (2/3NPYq-) (**B**) or from males lacking the entire Y chromosome long arm (NPYq-) (**C**) with (tsgic SP4) and without (neg sib) the addition of the *Flag-Sly1* SP4 transgene. Levels of protein expression were quantified with *ImageJ* software and normalized to Ponceau (PON) signal. The data represent an average $\pm$ SDev, with $n = 2$–3. Statistical significance (t-test, $p < 0.05$): bars marked with different letters are significantly different. D: Immunofluorescence detection of transgenic SLY1 protein (green) on stage VII testicular tubules from mice lacking the entire Y chromosome long arm without (NPYq-) and with (NPYq- tsgic) the *Flag-Sly1* SP4 transgene. Lectin (red) was used to stage the tubules and DAPI (blue) was used to stain the nuclei. Bar = 10 μm. (**E**) Sperm headshape analysis performed for NPYq-deficient mice without (2/3NPYq-) and with (2/3NPYq- tsgic) the *Flag-Sly1* SP4 transgene, and XY controls. Normal headshape and three categories of headshape defects were differentiated and quantified. The data represent an average $\pm$ SEM with $n = 13, 8$, and 3 males of the XY, 2/3NPYq-, and 2/3NPYq- tsgic genotype; a total of 1227, 651, and 274 sperm were examined for XY, 2/3NPYq-, and 2/3NPYq- tsgic genotype, respectively. Statistical significance: two-way ANOVA with genotype and sperm headshape as factors revealed a significant effect of genotype ($p = 0.0034$), headshape ($p < 0.0001$) and interaction ($p < 0.0001$). The results of paired comparison for specific sperm headshape between transgenic and negative siblings in post-hoc Bonferroni test are shown within graphs: [a] $p < 0.05$ when compared to XY and [b] $p < 0.05$ when compared to 2/3NPYq-.

4. Discussion

The goal of this study was to clarify whether the spermiogenic phenotype of mice with NPYq deficiencies was due to the absence of NPYq-encoded *Sly* gene. Prior analyses of mice with transgenically silenced *Sly* [11,12] provided strong evidence to support this hypothesis. However, one unresolved issue was whether *Sly* deficiency was the sole underlying cause. Here, we provide new data indicating that the spermiogenic phenotype in NPYq/*Sly* deficient mice correlates well with the deficiency of expression of endogenous *Sly*/SLY. However, transgenic rescue of *Sly*/SLY expression impairment in NPYq-deficient mice does not ameliorate their spermiogenic phenotype, suggesting that another NPYq encoded gene contributes to the spermiogenic phenotype.

4.1. New anti-SLY Antibody Confirms that Some SLY Protein is Retained in sh367 Sly-KD Mice

Sly (Sycp3-like Y-linked) is one of the genes encoded within a massively amplified region of the mouse Y chromosome; 126 copies of *Sly* with intact open reading frames are present on NPYq [1]. *Sly* encodes two main transcript variants, *Sly1* and *Sly2* [17]. *Sly1* is a full-length isoform and encodes a ~40-kDa protein SLY1 while *Sly2* originates from the alternative splicing of exons 5 and 6 and is translated to produce the shorter protein SLY2. Interestingly, exons 5 and 6 are duplicates of exons 3 and 4, and the functional difference (if any) between SLY1 and SLY2 remains unknown. Not all *Sly* copies on NPYq have the same structure and the picture is further complicated by additional copies expected to produce shorter ORF and some being non-coding.

The characterization of SLY protein(s) has to date focused on its most predominant isoform: SLY1. SLY1 has been shown to be very highly expressed in round spermatids, steps 2/3 to 9, with nuclear localization and, in particular, co-localization with sex chromosomes and *Speer* autosomal cluster during spermiogenesis [11,30]. The SLY protein contains a conserved COR1 domain initially identified in the synaptonemal complex protein SYCP3 and is able to interact with double-stranded DNA [31,32]. By chromatin immunoprecipitation, SLY1 protein has been shown to be enriched at the promoter of many spermatid-expressed genes, including spermatid-specific multicopy X and Y and genes involved in chromatin remodeling during spermiogenesis [14].

The sh367 Sly-KD mice were shown previously to have the most prominent *Sly* knockdown, sperm with gross head abnormalities, and severely impaired fertility [11]. These mice also have an increased incidence of DNA breaks in sperm and impaired sperm chromatin packaging [12]. The spermiogenic phenotype of sh367 mice was less pronounced than that of mice with severe NPYq deficiencies [11,12]. One of the intermediate *Sly* knockdown lines, line sh344, was shown to have moderate *Sly* transcript knockdown, no sperm headshape defects, good fertility, and no sperm DNA damage phenotype [12] While decreasing *Sly* transcript levels correlated well with the increasing severity of the phenotype, protein expression analyses were hampered by a lack of an appropriate antibody [30–32]. Here, using a new anti-SLY antibody allowing for the distinction between SLY1 and SLY2, we demonstrated that sh367 mice retain some SLY2, which could be responsible for their mild phenotype. We also demonstrated that phenotypically unaffected sh344 males lack most SLY1 but retain (and overexpress) SLY2, which emphasizes the role of the SLY2 isoform.

The fact that *Sly2* transcripts and protein are overexpressed when *Sly1* transcripts are knocked down is in agreement with observations that SLY negatively regulates its own expression [11,14]. Because the SLY2 sequence does not have anything unique compared to that of SLY1 and the alternatively spliced out region is in fact duplicated in SLY1, it is tempting to say that SLY1 and SLY2 have a similar function. The fact that *Sly* and NPYq- deficient phenotypes correlate well with *Sly1/2* transcript and SLY1/2 protein levels is further evidence in favor of this hypothesis (Table S3). Yet, only the production of an anti-SLY2 specific antibody, if feasible, will allow for confirmation of this assumption.

4.2. Transgenic SLY Rescues Sly/SLY Expression But Not the NPYq Specific Spermiogenic Phenotype, Suggesting the Involvement of Another NPYq Gene in the Same Pathway

Transgene rescue, along with gene knockout/knockdown, is a viable and commonly used approach for establishing gene function [33–36]. Therefore, to further elucidate whether the loss of *Sly* is solely responsible for the spermiogenic phenotype of mice with NPY deficiencies, we generated males transgenic for *Sly* and examined the effects of placing the transgene in the context of varying degrees of NPYq deficiencies ranging from 2/3NPYq to complete deletion of the NPYq region. In all the lines we produced, despite reaching a global level of *Sly1/2* transcripts and SLY1/2 proteins at least equal (and often superior) to what is observed in WT controls, we did not see any amelioration in the parameters we looked at: sperm morphology, number, motility, fertilizing ability, as well as post-meiotic sex chromatin (PMSC) gene expression.

What could be the reasons for the inability of transgenic SLY to overcome the spermiogenic phenotype of mice with NPYq deficiencies? One could suspect FLAG tag to interfere with the SLY protein function. However, immunoprecipitation and chromatin immunoprecipitation analyses performed with XY mice carrying the FLAG tagged transgenic SLY1 did not show any difference between FLAG-SLY1 transgenic protein and SLY1 endogenous protein in regard to binding partners and targets [14]. To further challenge this hypothesis, we generated mice transgenic for *Sly* but without the FLAG tag. The transgenic rescue with the *Sly* transgenes lacking the tag yielded similar results to the transgenes with *Flag-Sly* (Figure S8).

Another potential reason for the lack of spermiogenic phenotype rescue could be the timing of transgenic *Sly*/SLY expression. We selected the *SP10* promoter to drive the *Sly* transgenes in order to achieve a high spermatid-specific expression level from step 1 round spermatids [37,38]. As shown here and before [14], the *SP10* promoter drives the expression of transgenic *Sly* with a pattern very similar to that of endogenous *Sly*. Though slight differences may exist, it would more likely decrease the level of phenotypic rescue, rather than lead to a complete lack of it.

One can also ask whether the lack of phenotype rescue in lines carrying the *Flag-Sly1* transgene is due to a lack of expression rescue of *Sly2*/SLY2. Two pieces of evidence go against it. First, the results obtained with line 6P carrying both the *Flag-Sly1* and *Flag-Sly2* transgenes and showing both *Sly1* and *Sly2* transcripts and SLY1 and SLY2 protein expression rescue, presented no phenotype rescue. Second, the experiments performed with transgenic rescue mice carrying the *Sly1* and *Sly2* transgenes without *Flag*, showed that the addition of *Sly2* rescued the expression but did not rescue the phenotype (Figure S8).

All in all, we believe that the most likely reason for the lack of NPYq rescue with the *Sly* transgenes is that another NPYq encoded gene/s are involved in the same pathway. The most likely candidate is the multi-copy gene *Ssty*, which has been shown to be reduced in mice with NPY deficiencies [2]. On NPYq, 306 copies of *Ssty* with intact open reading frame have been identified [1]. Two versions of the *Ssty* gene (i.e., *Ssty1* and *Ssty2*) have been described and, as for *Sly*, a related spermatid-specific multicopy gene family exists on the X chromosome [39,40]. SSTY proteins belong to the SPIN family, all members of which bear three Spin/Ssty domains. One member of this family, SPIN1, has been extensively studied and was found to homo-dimerize and be able to recognize H3K4me3 and H3R8me2a [40,41].

SSTY molecular function has not been fully investigated, but we have shown that SSTY proteins are specifically expressed in spermatids, co-localize with PMSC, and interact with SLY and its X chromosome-linked homologues SLX/SLXL1 [42]. In a study in which H3K4me3 peptide was pulled down to identify chromatin readers and associated proteins, SSTY1, SSTY2, SLX/SLXL1, and SLY were found as part of the protein complexes interacting with H3K4me3 specifically in the testis [43], suggesting that SSTY proteins, like SPIN1, could bind to H3K4me3. As discussed in our previous studies [11,29], the SLY isoelectric point is acidic and, therefore, is not compatible for a direct interaction with DNA. Yet, ChIP-Seq analyses revealed that SLY1 genomic location overlaps with that of H3K4me3 in round spermatids [14]. A very tempting hypothesis is that SSTY proteins (possibly in the form of dimers) are required for SLY to interact with DNA/chromatin and control gene expression. According to this model, in a context in which both *Sly* and *Ssty* levels are decreased or absent, rescuing the sole expression of *Sly* would not improve the NPYq phenotype due to a lack of (enough) *Ssty*. Interestingly, the NPYq structure consists mostly of the repetition (in tandem or palindromic) of a ~500 kb unit containing one copy of *Sly* and three of *Ssty1/2* [1,44]. In the mouse lineage, they have been co-amplified [45], thus maintaining a similar stoichiometry between *Sly* and *Ssty* gene copies/amount of proteins.

Future work will involve testing for the *Ssty* roles by generating mice with *Ssty* knockdown as we have done with *Sly*. Transgene rescue of NPYq deficiency with *Ssty* transgenes should also be informative. We anticipate that combining the *Sly* and *Ssty* transgenic rescue might provide the most efficient reconstitution of normal spermatogenesis and fertility in a context of severe NPYq deficiency.

5. Conclusions

In conclusion, the transgenic rescue we used in this study is an interesting complementary approach to the gene knockout/knockdown strategy used before to investigate the NPYq gene function. Despite the previously established involvement of *Sly* in the NPYq- spermiogenic phenotype, a lack of rescue with the *Sly* transgene strongly suggests the participation of another NPYq gene. The best candidate identified to date is *Ssty*, which we suspect is required together with *Sly* for normal gene expression and chromatin regulation during spermiogenesis.

Supplementary Materials: The following are available online at http://www.mdpi.com/2073-4425/10/2/133/s1, Figure S1: Design of anti-SLY antibody, Figure S2: Characterization of anti-SLY antibody, Figure S3: Dot-blot analysis, Figure S4: SLY expression in males with NPYq- and Sly-specific deficiency, Figure S5: Production of mice transgenic for *Flag-Sly*, Figure S6: Addition of the *Flag-Sly* transgene to males with NPYq deletions rescues *Sly* expression deficiency, Figure S7: Addition of the *Flag-Sly* transgene to 2/3NPYq-males rescues SLY1 expression deficiency, Figure S8: Addition of the *Sly* transgene (no FLAG tag) to males with severe NPYq deficiency rescues *Sly* expression but not low sperm number and sperm ability to fertilize oocytes in vitro, Table S1: Summary of mice used in this study, Table S2: Primers, Table S3: Relationship between SLY1/2 protein expression and spermiogenic phenotype and fertility of mice with NPY/Sly deficiency.

Author Contributions: Conceptualization, project administration, supervision, data curation, formal analysis, writing, and funding acquisition, M.A.W. and J.C. Investigation, J.M.R., Y.Y., V.A.R., Q.U.M., M.B. and C.M.

Funding: This research was funded by NIH, HD072380, NIH RR024206, and Hawaii Community Foundation HCF13ADVC-60314 grants to M.A.W and by INSERM, FP7-PEOPLE-2010-IEF-273143 and ANR-17-CE12-0004-01 to J.C.

Acknowledgments: The authors wish to thank Paul Burgoyne for initiating the study of NPYq genes and supporting the present project. The authors are grateful to numerous undergraduate student volunteers in Ward lab who helped with mouse genotyping. The anti-SLY1/2 antibody was developed with help of the Monoclonal Antibody Service Facility and Training Center (MASFTC) at Kapiolani Community College; we are grateful to its Director John Berestecky and his staff. We also thank Aine Rattigan, Andrew Ojarikre, and other members of the Burgoyne lab at the NIMR for help with genotyping and breeding of SP4 *Flag-Sly1* mice and helpful discussions. We also thank Antoine Gueraud, members of the Mouse House Facility of the Cochin Institute, in particular Matthieu Benard and Amandine Janvier, and of the Cellular Imaging Facility of the Cochin Institute (INSERM U1016, CNRS UMR8104, Université Paris Descartes).

Conflicts of Interest: The authors declare no conflict of interest.

References

1. Soh, Y.Q.; Alfoldi, J.; Pyntikova, T.; Brown, L.G.; Graves, T.; Minx, P.J.; Fulton, R.S.; Kremitzki, C.; Koutseva, N.; Mueller, J.L.; et al. Sequencing the mouse y chromosome reveals convergent gene acquisition and amplification on both sex chromosomes. *Cell* **2014**, *159*, 800–813. [CrossRef] [PubMed]

2. Toure, A.; Clemente, E.J.; Ellis, P.; Mahadevaiah, S.K.; Ojarikre, O.A.; Ball, P.A.; Reynard, L.; Loveland, K.L.; Burgoyne, P.S.; Affara, N.A. Identification of novel Y chromosome encoded transcripts by testis transcriptome analysis of mice with deletions of the Y chromosome long arm. *Genome Biol.* **2005**, *6*, R102. [CrossRef]

3. Burgoyne, P.S.; Mahadevaiah, S.K.; Sutcliffe, M.J.; Palmer, S.J. Fertility in mice requires X-Y pairing and a Y-chromosomal "spermiogenesis" gene mapping to the long arm. *Cell* **1992**, *71*, 391–398. [CrossRef]

4. Moriwaki, K.; Suh, D.S. Genetic factors affecting sperm morphology in the mouse. *Mouse Newsl.* **1988**, *82*, 138.

5. Styrna, J.; Imai, H.T.; Moriwaki, K. An increased level of sperm abnormalities in mice with a partial deletion of the Y chromosome. *Genet. Res.* **1991**, *57*, 195–199. [CrossRef] [PubMed]

6. Styrna, J.; Klag, J.; Moriwaki, K. Influence of partial deletion of the Y chromosome on mouse sperm phenotype. *J. Reprod. Fertil.* **1991**, *92*, 187–195. [CrossRef] [PubMed]

7. Toure, A.; Szot, M.; Mahadevaiah, S.K.; Rattigan, A.; Ojarikre, O.A.; Burgoyne, P.S. A new deletion of the mouse Y chromosome long arm associated with the loss of Ssty expression, abnormal sperm development and sterility. *Genetics* **2004**, *166*, 901–912. [CrossRef]

8. Yamauchi, Y.; Riel, J.M.; Wong, S.J.; Ojarikre, O.A.; Burgoyne, P.S.; Ward, M.A. Live offspring from mice lacking the Y chromosome long arm gene complement. *Biol. Reprod.* **2009**, *81*, 353–361. [CrossRef]

9. Ward, M.A.; Burgoyne, P.S. The effects of deletions of the mouse Y chromosome long arm on sperm function—Intracytoplasmic sperm injection (ICSI)-based analysis. *Biol. Reprod.* **2006**, *74*, 652–658. [CrossRef]

10. Yamauchi, Y.; Riel, J.M.; Stoytcheva, Z.; Burgoyne, P.S.; Ward, M.A. Deficiency in mouse Y chromosome long arm gene complement is associated with sperm DNA damage. *Genome Biol.* **2010**, *11*, R66. [CrossRef]

11. Cocquet, J.; Ellis, P.J.; Yamauchi, Y.; Mahadevaiah, S.K.; Affara, N.A.; Ward, M.A.; Burgoyne, P.S. The multicopy gene Sly represses the sex chromosomes in the male mouse germline after meiosis. *PLoS Biol.* **2009**, *7*, e1000244. [CrossRef] [PubMed]

12. Riel, J.M.; Yamauchi, Y.; Sugawara, A.; Li, H.Y.; Ruthig, V.; Stoytcheva, Z.; Ellis, P.J.; Cocquet, J.; Ward, M.A. Deficiency of the multi-copy mouse Y gene Sly causes sperm DNA damage and abnormal chromatin packaging. *J. Cell Sci.* **2013**, *126*, 803–813. [CrossRef] [PubMed]

13. Ellis, P.J.; Clemente, E.J.; Ball, P.; Toure, A.; Ferguson, L.; Turner, J.M.; Loveland, K.L.; Affara, N.A.; Burgoyne, P.S. Deletions on mouse Yq lead to upregulation of multiple X- and Y-linked transcripts in spermatids. *Hum. Mol. Genet.* **2005**, *14*, 2705–2715. [CrossRef] [PubMed]

14. Moretti, C.; Serrentino, M.E.; Ialy-Radio, C.; Delessard, M.; Soboleva, T.A.; Tores, F.; Leduc, M.; Nitschke, P.; Drevet, J.R.; Tremethick, D.J.; et al. SLY regulates genes involved in chromatin remodeling and interacts with TBL1XR1 during sperm differentiation. *Cell Death Differ.* **2017**, *24*, 1029–1044. [CrossRef] [PubMed]

15. Reynard, L.N.; Turner, J.M. Increased sex chromosome expression and epigenetic abnormalities in spermatids from male mice with Y chromosome deletions. *J. Cell Sci.* **2009**, *122*, 4239–4248. [CrossRef] [PubMed]

16. Lavery, R.; Chassot, A.A.; Pauper, E.; Gregoire, E.P.; Klopfenstein, M.; de Rooij, D.G.; Mark, M.; Schedl, A.; Ghyselinck, N.B.; Chaboissier, M.C. Testicular differentiation occurs in absence of R-spondin1 and Sox9 in mouse sex reversals. *PLoS Genet.* **2012**, *8*, e1003170. [CrossRef] [PubMed]

17. Reynard, L.N.; Cocquet, J.; Burgoyne, P.S. The multi-copy mouse gene Sycp3-like Y-linked (Sly) encodes an abundant spermatid protein that interacts with a histone acetyltransferase and an acrosomal protein. *Biol. Reprod.* **2009**, *81*, 250–257. [CrossRef]

18. Akerfelt, M.; Henriksson, E.; Laiho, A.; Vihervaara, A.; Rautoma, K.; Kotaja, N.; Sistonen, L. Promoter ChIP-chip analysis in mouse testis reveals Y chromosome occupancy by HSF2. *Proc. Natl. Acad. Sci. USA* **2008**, *105*, 11224–11229. [CrossRef]

19. Garcia, M.A.; Collado, M.; Munoz-Fontela, C.; Matheu, A.; Marcos-Villar, L.; Arroyo, J.; Esteban, M.; Serrano, M.; Rivas, C. Antiviral action of the tumor suppressor ARF. *EMBO J.* **2006**, *25*, 4284–4292. [CrossRef]

20. Kimura, Y.; Yanagimachi, R. Intracytoplasmic sperm injection in the mouse. *Biol. Reprod.* **1995**, *52*, 709–720. [CrossRef]

21. Chatot, C.L.; Ziomek, C.A.; Bavister, B.D.; Lewis, J.L.; Torres, I. An improved culture medium supports development of random-bred 1-cell mouse embryos in vitro. *J. Reprod. Fertil.* **1989**, *86*, 679–688. [CrossRef] [PubMed]

22. Quinn, P.; Barros, C.; Whittingham, D.G. Preservation of hamster oocytes to assay the fertilizing capacity of human spermatozoa. *J. Reprod. Fertil.* **1982**, *66*, 161–168. [CrossRef] [PubMed]

23. Ajduk, A.; Yamauchi, Y.; Ward, M.A. Sperm chromatin remodeling after intracytoplasmic sperm injection differs from that of in vitro fertilization. *Biol. Reprod.* **2006**, *75*, 442–451. [CrossRef] [PubMed]

24. Ward, M.A.; Yanagimachi, R. Intracytoplasmic Sperm Injection in Mice. *Cold Spring Harb. Protoc.* **2018**, *2018*, pdb.prot094482. [CrossRef] [PubMed]

25. Burgoyne, P.S.; Mahadevaiah, S.K.; Perry, J.; Palmer, S.J.; Ashworth, A. The Y* rearrangement in mice: New insights into a perplexing PAR. *Cytogenet. Cell Genet.* **1998**, *80*, 37–40. [CrossRef]

26. Cattanach, B.M. Sex-reversed mice and sex determination. *Ann. N. Y. Acad. Sci.* **1987**, *513*, 27–39. [CrossRef] [PubMed]

27. Cattanach, B.M.; Pollard, C.E.; Hawker, S.G. Sex-reversed mice: XX and XO males. *Cytogenetics* **1971**, *10*, 318–337. [CrossRef]

28. Eicher, E.M.; Hale, D.W.; Hunt, P.A.; Lee, B.K.; Tucker, P.K.; King, T.R.; Eppig, J.T.; Washburn, L.L. The mouse Y* chromosome involves a complex rearrangement, including interstitial positioning of the pseudoautosomal region. *Cytogenet. Cell Genet.* **1991**, *57*, 221–230. [CrossRef]

29. Comptour, A.; Moretti, C.; Serrentino, M.E.; Auer, J.; Ialy-Radio, C.; Ward, M.A.; Toure, A.; Vaiman, D.; Cocquet, J. SSTY proteins co-localize with the post-meiotic sex chromatin and interact with regulators of its expression. *FEBS J.* **2014**, *281*, 1571–1584. [CrossRef]

30. Cocquet, J.; Ellis, P.J.; Mahadevaiah, S.K.; Affara, N.A.; Vaiman, D.; Burgoyne, P.S. A genetic basis for a postmeiotic X versus Y chromosome intragenomic conflict in the mouse. *PLoS Genet.* **2012**, *8*, e1002900. [CrossRef]

31. Syrjanen, J.L.; Pellegrini, L.; Davies, O.R. A molecular model for the role of SYCP3 in meiotic chromosome organisation. *eLife* **2014**, *3*. [CrossRef] [PubMed]

32. Yuan, L.; Liu, J.G.; Zhao, J.; Brundell, E.; Daneholt, B.; Hoog, C. The murine SCP3 gene is required for synaptonemal complex assembly, chromosome synapsis, and male fertility. *Mol. Cell* **2000**, *5*, 73–83. [CrossRef]

33. Mazeyrat, S.; Saut, N.; Grigoriev, V.; Mahadevaiah, S.K.; Ojarikre, O.A.; Rattigan, A.; Bishop, C.; Eicher, E.M.; Mitchell, M.J.; Burgoyne, P.S. A Y-encoded subunit of the translation initiation factor Eif2 is essential for mouse spermatogenesis. *Nat. Genet.* **2001**, *29*, 49–53. [CrossRef] [PubMed]

34. Vernet, N.; Mahadevaiah, S.K.; Ellis, P.J.; de Rooij, D.G.; Burgoyne, P.S. Spermatid development in XO male mice with varying Y chromosome short-arm gene content: evidence for a Y gene controlling the initiation of sperm morphogenesis. *Reproduction* **2012**, *144*, 433–445. [CrossRef] [PubMed]

35. Yamauchi, Y.; Riel, J.M.; Ruthig, V.A.; Ortega, E.A.; Mitchell, M.J.; Ward, M.A. Two genes substitute for the mouse Y chromosome for spermatogenesis and reproduction. *Science* **2016**, *351*, 514–516. [CrossRef] [PubMed]

36. Yamauchi, Y.; Riel, J.M.; Stoytcheva, Z.; Ward, M.A. Two Y genes can replace the entire Y chromosome for assisted reproduction in the mouse. *Science* **2014**, *343*, 69–72. [CrossRef]

37. Reddi, P.P.; Shore, A.N.; Shapiro, J.A.; Anderson, A.; Stoler, M.H.; Acharya, K.K. Spermatid-specific promoter of the SP-10 gene functions as an insulator in somatic cells. *Dev. Biol.* **2003**, *262*, 173–182. [CrossRef]

38. Reddi, P.P.; Flickinger, C.J.; Herr, J.C. Round spermatid-specific transcription of the mouse SP-10 gene is mediated by a 294-base pair proximal promoter. *Biol. Reprod.* **1999**, *61*, 1256–1266. [CrossRef]

39. Laval, S.H.; Reed, V.; Blair, H.J.; Boyd, Y. The structure of DXF34, a human X-linked sequence family with homology to a transcribed mouse Y-linked repeat. *Mamm. Genome* **1997**, *8*, 689–691. [CrossRef]

40. Oh, B.; Hwang, S.Y.; Solter, D.; Knowles, B.B. Spindlin, a major maternal transcript expressed in the mouse during the transition from oocyte to embryo. *Development* **1997**, *124*, 493–503.

41. Su, X.; Zhu, G.; Ding, X.; Lee, S.Y.; Dou, Y.; Zhu, B.; Wu, W.; Li, H. Molecular basis underlying histone H3 lysine-arginine methylation pattern readout by Spin/Ssty repeats of Spindlin1. *Genes Dev.* **2014**, *28*, 622–636. [CrossRef] [PubMed]

42. Toure, A.; Grigoriev, V.; Mahadevaiah, S.K.; Rattigan, A.; Ojarikre, O.A.; Burgoyne, P.S. A protein encoded by a member of the multicopy Ssty gene family located on the long arm of the mouse Y chromosome is expressed during sperm development. *Genomics* **2004**, *83*, 140–147. [CrossRef]

43. Eberl, H.C.; Spruijt, C.G.; Kelstrup, C.D.; Vermeulen, M.; Mann, M. A map of general and specialized chromatin readers in mouse tissues generated by label-free interaction proteomics. *Mol. Cell* **2013**, *49*, 368–378. [CrossRef] [PubMed]

44. Ellis, P.J.; Ferguson, L.; Clemente, E.J.; Affara, N.A. Bidirectional transcription of a novel chimeric gene mapping to mouse chromosome Yq. *BMC Evol. Biol.* **2007**, *7*, 171. [CrossRef] [PubMed]

45. Ellis, P.J.; Bacon, J.; Affara, N.A. Association of Sly with sex-linked gene amplification during mouse evolution: A side effect of genomic conflict in spermatids? *Hum. Mol. Genet.* **2011**, *20*, 3010–3021. [CrossRef] [PubMed]

 *genes*

Article

A Comparison Between Two Assays for Measuring Seminal Oxidative Stress and their Relationship with Sperm DNA Fragmentation and Semen Parameters

Sheryl T. Homa [1,2,*], Anna M. Vassiliou [1,2], Jesse Stone [1], Aideen P. Killeen [2], Andrew Dawkins [2], Jingyi Xie [2], Farley Gould [2] and Jonathan W. A. Ramsay [3]

[1]   Department of Biosciences, University of Kent, Canterbury CT2 7NJ, UK;
      anna@andrologysolutions.co.uk (A.M.V.); jlstone19@gmail.com (J.S.)
[2]   Department of Andrology, The Doctors Laboratory, London W1G 9RT, UK;
      aideen.killeen@tdlpathology.com (A.P.K.); andrew.dawkins@tdlpathology.com (A.D.);
      jingyi-xie@hotmail.co.uk (J.X.); FGould@hotmail.co.uk (F.G.)
[3]   Imperial College Healthcare NHS Trust, London W2 1NY, UK; jonathan.ramsay@imperial.nhs.uk
*   Correspondence: s.homa@andrologysolutions.co.uk; Tel.: +44-(0)-20-7224-2322

Received: 18 January 2019; Accepted: 11 March 2019; Published: 19 March 2019

Abstract: Oxidative stress (OS) is a significant cause of DNA fragmentation and is associated with poor embryo development and recurrent miscarriage. The aim of this study was to compare two different methods for assessing seminal OS and their ability to predict sperm DNA fragmentation and abnormal semen parameters. Semen samples were collected from 520 men attending for routine diagnostic testing following informed consent. Oxidative stress was assessed using either a chemiluminescence assay to measure reactive oxygen species (ROS) or an electrochemical assay to measure oxidation reduction potential (sORP). Sperm DNA fragmentation (DFI) and sperm with immature chromatin (HDS) were assessed using sperm chromatin structure assay (SCSA). Semen analysis was performed according to WHO 2010 guidelines. Reactive oxygen species sORP and DFI are negatively correlated with sperm motility (p = 0.0012, 0.0002, <0.0001 respectively) and vitality (p < 0.0001, 0.019, <0.0001 respectively). The correlation was stronger for sORP than ROS. Reactive oxygen species (p < 0.0001), sORP (p < 0.0001), DFI (p < 0.0089) and HDS (p < 0.0001) were significantly elevated in samples with abnormal semen parameters, compared to those with normal parameters. Samples with polymorphonuclear leukocytes (PMN) have excessive ROS levels compared to those without (p < 0.0001), but sORP and DFI in this group are not significantly increased. DNA fragmentation was significantly elevated in samples with OS measured by ROS (p = 0.0052) or sORP (p = 0.004). The results demonstrate the multi-dimensional nature of oxidative stress and that neither assay can be used alone in the diagnosis of OS, especially in cases of leukocytospermia.

Keywords: oxidative stress; reactive oxygen species; chromatin; DNA fragmentation; DNA oxidation; male infertility; spermatogenesis

---

1. Introduction

Oxidative stress (OS) is thought to be the pathologic molecular mechanism underpinning the majority of known clinical, environmental and lifestyle causes of male infertility. It is associated with varicocoele, genitourinary tract infection, prostatitis, obesity, tobacco smoking, endocrine imbalance and testicular dysfunction [1–4]. OS occurs when the physiological balance of oxidants and reductants in a system is disturbed as a result of excessive production of reactive oxygen species (ROS) or a reduction in levels of antioxidants [5]. Oxidative balance is essential for normal sperm function [2,4,6–9]. Reactive oxygen species include superoxide anion ($O_2^{-\bullet}$), hydroxyl radical

(OH•), hydrogen peroxide (H_2O_2), peroxynitrite (ONOO−), nitric oxide (•NO) or hypochlorous acid (HOCl). Reactive oxygen species are required at low levels for chromatin and flagellar protein modification during spermatogenesis [10] as well as for the normal process of sperm hyperactivation and capacitation [2,6–9]. However, at high levels, OS impairs fertilisation through interference with capacitation and the acrosome reaction [4,6,8].

Sperm are exposed to OS during spermatogenesis as well as during epididymal storage and transit through the reproductive tract and at ejaculation [4,8,11]. Reactive oxygen species may be produced extrinsically by infiltrating polymorphonuclear leukocytes (PMN) [12–15] or from the presence of oxidants in the seminal fluid resulting from tobacco smoking, excessive testicular heat or other environmental toxins [3,4]. On the other hand, ROS can be generated intrinsically, primarily as a result of electron leakage in the sperm mitochondria, from cytosolic L-amino acid oxidases and plasma membrane nicotinamide adenine dinucleotide phosphate (NADP) oxidases [6,16,17]. High levels of ROS are also produced by abnormal sperm that retain excess residual cytoplasm as a result of incomplete sperm maturation [6,18]. As the sperm have negligible sources of intracellular antioxidants, ROS levels may remain elevated, leaving vulnerable molecules susceptible to oxidative attack [6,8]. Reactive oxygen species are extremely harmful because they target every cellular constituent, which has serious consequences for cell signalling and the function of the sperm. The sperm plasma membrane is particularly susceptible to oxidation as it is enriched in polyunsaturated fatty acids. These lipids can be oxidised through a series of chain reactions to release potentially toxic and mutagenic aldehydes and alkenals [6,15,19,20]. Importantly, sperm DNA is exquisitely sensitive to oxidative attack, resulting in impairment of embryo development, increased risk of gene mutations and miscarriage, congenital malformations and a high frequency of diseases in the offspring [3,5,8,11,19,21,22]. A serious consequence of OS is that it interferes with epigenetic modification and there are reports of abnormalities in sperm gene methylation as a direct result of oxidative insult [23–25]. There is good evidence to link unexplained infertility and recurrent pregnancy loss with both oxidative stress and sperm DNA fragmentation which are significantly elevated in infertile men [26–29] and in men whose partners experience miscarriage [11,30,31].

For many decades, semen analysis has been considered the gold standard for assessment of male infertility. However, this subjective microscopic analysis is poorly correlated with infertility and fails to provide any information about sperm function. More recently, assessment of sperm DNA fragmentation has been implemented as a more reliable marker for male infertility [32], yet it also has limitations as it does not address the plethora of other physiological and pathological functions regulated by oxidative stress in sperm. Assessment of OS can be performed in semen using a chemiluminescence assay [33–35]. This test measures the oxidation of luminol, a chemiluminescent probe, providing information about the levels of oxidants in the system. Alternatively, oxidative stress can be measured using a novel electrochemical assay which determines the oxidation reduction potential of the system taking into consideration all of the oxidants and antioxidants that are present [36,37]. The latter test is a simpler and more efficient method to determine OS. However, the two tests look at different aspects of oxidative stress, but it is precisely because of this that we consider it important to determine which assay may be more clinically relevant if we are to implement them as useful diagnostic tests. In order for an assay to be accepted as a clinically relevant diagnostic test, it should be able to predict abnormalities in markers known to be affected by it. While previous studies of the chemiluminescence and redox assays have shown an association of OS with clinical markers such as semen parameters [28,38–41] and sperm DNA fragmentation [40,42–46], to date a comparison of the two assays has not been performed. This study presents the first comparison of the two assays to determine which assay is more predictive of impaired semen quality. The association of oxidative stress markers (ROS and sORP) with sperm DNA fragmentation and semen parameters was investigated. Given that leukocytes are a major source of extrinsic ROS and that the role of leukocytes in male infertility remains controversial, the assays were also compared between samples with and without leukocytospermia. The results show a clear association between OS and sperm DNA damage, as well as impaired semen parameters irrespective of the method used to measure OS, although sORP is more predictive than ROS, especially in cases of leukocytospermia. It is proposed that differences between the methods may be explained in part by a difference in sensitivity for measurement of OS in extrinsic versus intrinsic compartments of the sperm in seminal fluid.

2. Materials and Methods

2.1. Ethics Statement

This study was approved by the Faculty of Sciences Research Ethics Advisory Group for Human Participants at the University of Kent (ID number 0601516) and adhered to the current legislation on research involving human subjects in the UK.

2.2. Semen Samples

Semen samples were collected from men who were attending for diagnostic semen analysis and who had given their informed consent to use any of the sample that remained after analysis for the study. Participants were advised to have 2–5 days sexual abstinence before providing a sample on site. A total of 599 samples were provided for the study, of which 79 were excluded. Exclusion criteria were incomplete sample collection, febrile illness during the previous 12 weeks, both of which may have affected the reliability of the results, and samples containing less than 1 million/mL sperm as ROS and sORP measurement are inaccurate and unreliable when the sperm concentration falls below this value. Of the men who consented, 496 were included in the study. Of these, 24 had attended for a repeat semen analysis following a clinical management plan. Semen analysis was performed as part of diagnostic testing according to the WHO 2010 criteria [47]. Samples were incubated at 36°C (± 1 °C) to liquefy prior to analysis. All samples were analysed at 20 ± 5 min after production. Leukocytes were identified using a peroxidase screen (LeucoScreen, FertiPro N.V. Belgium) and differential cell counting on Papanicoloau stained slides assessed under oil immersion at $\times 1000$ magnification. The distribution of semen samples in the study cohorts is shown in Table 1. The heterogeneity of the semen samples is evenly distributed among the OS study groups. Teratozoospermia was the most prevalent abnormality in both study groups, constituting 24% of all samples. Less than 10% of samples assessed for OS had abnormalities in all three major semen parameters.

Table 1. Distribution of semen samples classified according to WHO (2010) criteria in the patient study cohorts.

	ROS		sORP	
	Number of Patients	**%**	**Number of Patients**	**%**
Normozoopspermia	172	34.6	139	46.2
Oligozoospermia	18	3.6	9	3.0
Asthenozoospermia	8	1.6	6	2.0
Teratozoospermia	119	24.0	72	23.9
Oligoasthenozoospermia	1	<1	1	<1
Oligoteratozoospermia	69	13.9	32	10.6
Asthenoteratozoospermia	29	5.8	11	3.7
Oligoasthenoteratozoospermia	49	9.9	15	5.0
Leukocytospermia	31	6.3	16	5.3
TOTAL	496	-	301	-

ROS = reactive oxygen species; sORP = static oxidation reduction potential.

2.3. MeasuringReactive Oxygen Species Using Chemiluminescence

Reactive oxygen specieslevels were measured using a CE-marked single-tube Luminometer (Modulus Model no. 9200-001; Turner Biosystems Instrument Inc., Sunnyvale, CA, USA). Luminol was used as the probe, which is oxidised in the presence of ROS, resulting in chemiluminescence. The general methodology for this test has been reviewed elsewhere [34,35]. Briefly, negative and positive controls were run daily. Negative controls were prepared using 400 μL phosphate buffered saline (PBS) with 10 μL of a luminol working solution (Sigma-Aldrich, Dorset, UK) (5 mM luminol prepared in dimethylsulphoxide (DMSO)). Positive controls were prepared using 395 μL PBS, 5 μL 30% H_2O_2 and 10 μL of 5 mM luminol working solution. When measuring ROS in semen, 10 μL of luminol working solution was added to 400 μL of liquefied whole semen at 20 min post-ejaculation and measured immediately. Results were normalized to the sperm concentration and reported in relative light units (RLU)/sec/10^6 sperm.

2.4. Measuring Oxidation-Reduction Potential Using MiOXSYS

The ORP of semen samples was measured using the CE-marked MiOXSYS platform (MiOXSYS, Aytu BioScience Inc., Englewood, CO, USA). The analyser consists of an ultrahigh impedance electrometer with a self-contained electrochemical cell with platinum working and counter electrodes, and a 3 M KCl, Ag/AgCl reference electrode [48]. For the assay, a 30 μL sample is applied to a sensor which is inserted into the analyser. The voltage is measured between the reference and working electrodes every 0.5 s. The final sORP (static oxidation reduction potential) reading on the analyser display screen is the average of the final 10 s (20 readings). Higher sORP readings indicate an imbalance that favours the pro-oxidants and therefore suggests the presence of oxidative stress in the sample [48]. A reading in mV is displayed on the MiOXSYS analyser. This value is normalised to the sperm concentration of the sample. The result is reported as mV/10^6 sperm/mL.

2.5. Measuring Sperm DNA Fragmentation

The sperm chromatin structure assay (SCSA®) was utilized to assess DNA fragmentation. Details of the SCSA® have been described in detail elsewhere [32]. In brief, sperm are treated with a low pH buffer for 30 sec that opens up the two DNA strands where there is either a single (sd) or double (ds) DNA strand break. Acridine orange complexes with ds DNA and fluoresces green while complexing with sd DNA produces red fluorescence. Those sperm with any measurable increase in red fluorescence are scored as sperm with DNA fragmentation (DFI). Additionally, we included assessment of the fraction of high DNA stainable (HDS) cells, which are considered to represent immature spermatozoa with incomplete chromatin condensation.

2.6. Statistical Analysis

All data were analysed using the Statistical Analysis Systems software package (SAS Inst. Inc., Cary, NC) version 9.4. Data were tested for adherence to normality using PROC UNIVARIATE (SAS, 2013). The CORR procedure of SAS (PROC CORR, SAS 2013) was used to determine correlations between various semen parameters. Pearson correlation coefficients (R^2) and *p* values were estimated and reported for all parameters. Due to the expected non-normality of quantitative variables in this study, group comparisons were performed with Kruskal–Wallis test for 3-group comparisons, or Wilcoxon rank sum test for pairwise group comparisons using the NPAR1WAY procedure of SAS (PROC NPAR1WAY, SAS 2013). These nonparametric tests were used for age, sperm concentration/mL, total motility, progressive motility, vitality, morphology, PMN concentration, ROS, sORP, DFI and HDS values. In all cases, *p* values < 0.05 were considered to be statistically significant.

3. Results

3.1. Correlation between OS and Sperm DNA Damage; Comparison between Two Methods of OS Measurement

This study investigated whether there is a direct correlation between OS and DNA damage and whether the observations are consistent between the two methods of OS measurement. Observations were made with and without inclusion of samples with leukocytospermia as they are known to generate high levels of exogenous ROS and may obscure the effects of ROS generated endogenously. Initially, it was necessary to determine whether detection of OS was comparable between the two methods of assessment. Oxidative stress was assessed in 315 samples using either the chemiluminescence assay or oxidation reduction potential assay. Results showed only a weak but nevertheless significant positive correlation between observations for ROS and sORP (R^2 = 0.1172, *p* = 0.0376, n = 315). Interestingly, when samples with leukocytospermia are excluded, the correlation between sORP and ROS is marginally stronger (R^2 = 0.15095, *p* = 0.0089, n = 299). When OS was compared to DFI levels, ROS was highly significantly correlated to DFI, exhibiting a moderate positive relationship (R^2 = 0.24316, *p* = 0.0002, n = 237). Oxidation reduction potential shows a similar relationship to DFI, however this is not significant and may be due to the relatively low sample numbers (R^2 = 0.23992, *p* = 0.1043, n = 47). The ROS versus DFI correlation is also slightly stronger in the absence of data from patients exhibiting elevated PMN (R^2 = 0.31139, *p* < 0.0001, n = 222), however this is not the case for sORP. This is likely because only one patient from the sORP group exhibited leukocytospermia (R^2 = 0.22706, *p* = 0.1291, n = 46). In contrast, HDS shows no significant correlation with oxidative stress, whether it is measured by ROS (R^2 = 0.11211, *p* = 0.085, n = 237) or sORP (R^2 = 0.01222 *p* = 0.9351, n = 47), irrespective of leukocytospermia (excluding PMN: ROS R^2 = 0.10329, *p* = 0.1249, n = 222; sORP R^2 = 0.01853, *p* = 0.9027, n = 46).

3.2. Sperm DNA Damage and HDS Levels in Oxidative Balanced versus Oxidative Stressed Semen Samples

The chemiluminescence and MiOXSYS assays have both been validated and verified in-house at The Doctors Laboratory, which is ISO15189 UKAS accredited. The reference ranges determined by ROC analysis were ≤13.8 RLU/sec/10^6 sperm/mL (86% sensitivity; 86% specificity) for ROS and ≤1.4 mV/10^6 sperm/mL (76.4% sensitivity; 75.9% specificity) for sORP. Samples are considered to be in oxidative stress if they exceed the clinical reference values. When the patient cohort is separated into groups with or without oxidative stress, mean DFI (Table 2) was significantly elevated in the OS group irrespective of the method of OS measurement used, although the difference was much more significant when OS was measured by redox potential. When samples with PMN are excluded, the DFI is slightly higher in the group with OS as measured by ROS (24.67 ± 1.78 vs. 22.86 ± 1.59), but this difference was not significant. There were no samples with leukocytospermia in the group with OS as measured by sORP. While %HDS was also significantly elevated in samples with high levels of ROS (Table 3), the increase seen in %HDS in samples with elevated sORP was not significant, which may be attributed to the lower numbers of samples in this test group.

Table 2. Sperm DNA fragmentation in the presence and absence of oxidative stress.

	All Samples	All Samples Excluding those with Leukocytospermia
Oxidative stress measured by ROS		
Oxidative balanced	18.78 ± 1.10 (161)	18.75 ± 1.12 (159)
Oxidative stressed	22.86 ± 1.59 (77)	24.67 ± 1.78 (63)
p value	0.0359	0.0052
Oxidative stress measured by sORP		
Oxidative balanced	11.97 ± 1.41 (30)	12.14 ± 1.49 (28)
Oxidative stressed	19.39 ± 1.83 (18)	19.39 ± 1.86 (18)
p value	0.0024	0.004

Values represent the mean %DFI ± SEM. Number of samples in parentheses.

Table 3. High DNA stainability of sperm in the presence and absence of oxidative stress.

	All Samples	All Samples Excluding those with Leukocytospermia
Oxidative stress measured by ROS		
Oxidative balanced	13.45 ± 0.74 (161)	13.49 ± 0.75 (159)
Oxidative stressed	15.78 ± 1.02 (77)	16.19 ± 1.16 (63)
p value	0.0097	0.0077
Oxidative stress measured by sORP		
Oxidative balanced	11.07 ± 1.11 (30)	10.61 ± 1.09 (28)
Oxidative stressed	17.89 ± 3.40 (18)	17.89 ± 3.40 (18)
I value	0.0881	0.0672

Values represent the mean %HDS $\pm$ SEM. Number of samples in parentheses.

3.3. Correlation between Oxidative Stress, Sperm DNA Damage and Semen Parameters

Oxidative stress is manifested in poor semen quality. Using the two different methods for measuring OS, the results demonstrate a highly significant negative correlation between OS and total motility, progressive motility, total motile sperm count, vitality and morphology (see Table 4). The correlation is approximately twice as strong when OS is measured by sORP compared to ROS for all parameters with the exception of vitality which shows a stronger and more significant correlation with ROS (sORP: $R^2 = -0.13519$, $p = 0.019$; ROS: $R^2 = -0.20832$, $p < 0.0001$). This indicates that measurement of sORP may be a more sensitive marker for oxidative stress than ROS. An even stronger, highly significant negative correlation is seen between DFI and semen parameters, particularly with total ($R^2 = -0.53951$, $p = {<}0.0001$) and progressive motility ($R^2 = -0.48693$, $p < 0.0001$) and vitality ($R^2 = -0.5727$, $p < 0.0001$).

Table 4. Correlation between oxidative stress, sperm genetic integrity and semen parameters.

	Value	Count/ml	Total Motility	Progressive Motility	Total Motile Sperm Count	Vitality	Morphology
ROS	R^2	−0.15729	−0.14482	−0.14444	−0.17395	−0.20832	−0.12536
	p value	0.0004	0.0012	0.0013	0.0001	<0.0001	0.0053
	n	496	495	495	495	495	493
sORP	R^2	−0.24628	−0.21101	−0.23561	−0.25055	−0.13519	−0.22642
	p value	<0.0001	0.0002	<0.0001	<0.0001	0.019	<0.0001
	n	301	301	301	301	301	300
DFI	R^2	-0.19182	−0.53951	−0.48693	−0.27539	−0.5727	−0.19016
	p value	0.0041	<0.0001	<0.0001	<0.0001	<0.0001	0.0047
	n	222	221	221	221	221	220
HDS	R^2	−0.36663	−0.23638	−0.24938	−0.27703	−0.11497	−0.48848
	p value	<0.0001	0.0004	0.0002	<0.0001	0.0882	<0.0001
	n	222	221	221	221	221	220
PMN	R^2	0.2098	0.03497	0.04169	0.15498	0.03037	0.04361
	p value	<0.0001	0.4389	0.3561	0.0006	0.5015	0.3354
	n	493	492	492	492	492	490

R^2 = Pearson correlation coefficients; n = number of samples. ROS = Reactive oxygen species; sORP = oxidation-reduction potential; DFI = DNA fragmentation index; HDS = sperm with immature chromatin; PMN = polymorphonuclear leukocytes.

In contrast, HDS levels are not correlated with vitality, but are negatively correlated with all other semen parameters. The strongest correlation is between HDS and morphology and is highly significant ($R^2 = -0.48848$, $p = 0 < 0.0001$). Oxidation reduction potential is also highly significantly negatively correlated to morphology ($R^2 = -0.22642$, $p = {<}0.0001$), although not as strong a correlation as between HDS and morphology. Polymorphonuclear leukocytes are known to produce high levels of ROS, however presence of PMN in seminal fluid shows no correlation with classical markers of oxidative damage in sperm, including motility, vitality and DNA damage (see Table 4), although there is a strong positive correlation with sperm count.

3.4. Comparison of Sperm DNA Damage and OS among Different Patient Groups Selected According to Semen Parameters

To further evaluate the correlation between OS, DNA damage and semen parameters, patients were grouped according to whether they had normal or abnormal semen parameters. As PMN are well known to generate high levels of ROS, samples containing $\geq 1 \times 10^6$ million/ml PMN were grouped in a separate category. Reactive oxygen species, sORP, DFI and HDS levels were analysed between the different patient groups. Figure 1 shows OS levels are significantly higher in semen samples with one or more abnormal semen parameters compared to samples with normal semen parameters as expected, irrespective of the method of OS measurement (Figure 1a ROS: $p < 0.001$; Fig 1b sORP: $p < 0.007$). Median ROS levels are 0.80 (range 0–319.6) for normal semen samples versus 2.95 (range 0–1755) for abnormal samples, while median sORP is 0.44 (range -0.18–18.16) for normal samples versus 1.31 (range -0.78–59.43) for abnormal samples.

Reactive oxygen species levels are highest in the group of men with leukocytospermia and are significantly higher than in men with normal semen parameters (PMN: median 71.3, range 0.9–957.2 vs. normal: 0.8, range 0–319.6) ($p < 0.0001$). Unexpectedly, unlike the results observed with ROS measurement, sORP levels are not significantly different between the group with normal semen parameters and the group with PMN (Figure 1b sORP normal: median 0.44, range -0.18–18.16 vs. PMN: median 0.40, range 0.06–1.49).

Figure 2 shows the difference in sperm DNA damage between patient groups. While median DFI is significantly higher in samples with abnormal semen parameters compared to those with normal parameters (Figure 2a: median 18; range 2–81 vs. median 11; range 3–48 respectively; $p = 0.0089$), as with sORP, median DFI in the leukocytospermia samples (10; range 3–32) is not significantly different to those with normal semen parameters.

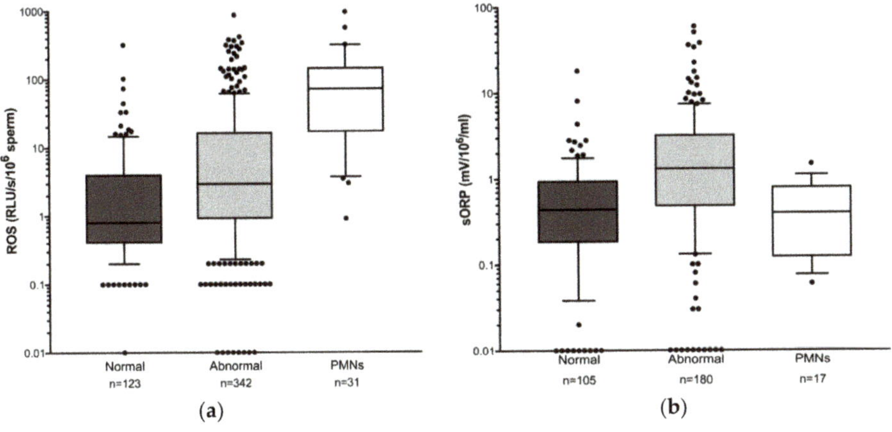

Figure 1. Box and whisker plots for seminal oxidative stress levels in different patient groups showing the median and interquartile ranges. Normal—normal semen parameters; Abnormal—abnormal parameters with <1 million/mL leukocytes; PMN—any parameters with $\geq$1 million/mL leukocytes. OS was measured using either (**a**) chemiluminescence (ROS) or (**b**) oxidation reduction potential (sORP) Lower whisker = 10th percentile; upper whisker = 90th percentile. Dots indicate values outside the range. Data are shown on a logarithmic scale.

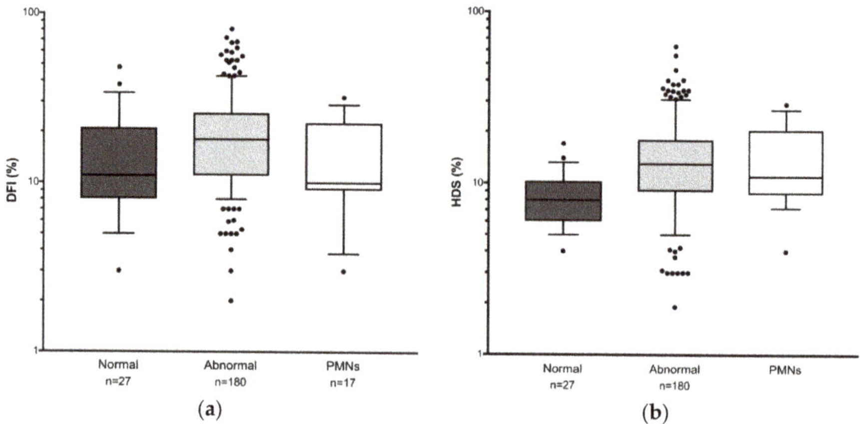

Figure 2. Box and whisker plots for sperm genetic integrity in different patient groups showing the median and interquartile ranges. Normal—normal semen parameters; Abnormal—abnormal parameters with <1 million/mL leukocytes; PMN—any parameters with ≥1 million/mL leukocytes. Sperm DNA fragmentation and HDS were assessed in the same samples. (**a**) DNA fragmentation (DFI) (**b**) immature chromatin (HDS). Lower whisker = 10th percentile; upper whisker = 90th percentile. Dots indicate values outside the range. Data are shown on a logarithmic scale.

Interestingly, the mean age of men in the PMN group is significantly higher than in the other two groups (PMN 41.50 ± 1.34 vs. normal 38.46 ± 0.57, $p = 0.0342$; vs. abnormal 38.25 ± 0.34, $p = 0.0134$). As it has previously been reported that sperm DNA becomes increasingly fragmented as men age [3,49], one might expect that samples in the leukocytospermia group would exhibit the highest levels of sperm DNA damage, but clearly this is not the case. On the other hand, HDS levels are significantly raised in both the leukocytospermic and abnormal semen parameter groups compared to samples with normal semen parameters (Figure 2b; normal: median 8; range 4–17 vs. abnormal: median 13; range 1.9–63.1 $p = < 0.0001$; vs. PMN: median 11; range 4–29 $p = 0.0075$) with no overlap between the boxes, suggesting a different mechanism of action for sperm DNA damage and sperm genetic maturation.

4. Discussion

A systematic review of the literature comparing different methods for measuring OS in terms of the practicality of the methodology of the tests in the laboratory, cost effectiveness and sensitivity and specificity of the tests has revealed the superiority of the MiOXSYS assay [50]. Our study expands on the review by Agarwal et al. [50] as it is the first to perform a comparison of two different tests for measuring oxidative stress (chemiluminescence versus redox potential) in terms of their efficacy in predicting impaired semen parameters and sperm chromatin structure, and hence their usefulness as a clinical diagnostic test. Novel findings from our study demonstrate that redox potential is more highly correlated with semen parameters than measurement of ROS. Furthermore, established clinical reference ranges for OS tests are extremely useful in categorising patients with altered sperm chromatin structure or sperm DNA damage, when the DFI threshold is 25%. We have shown for the first time that this is particularly more evident for those patients exhibiting OS as measured by redox potential. This study also investigated the ambiguity surrounding the measurement of oxidative stress in leukocytospermia. We compared oxidative stress measured by chemiluminescence versus oxidation reduction potential in the presence and absence of leukocytospermia. The data we report for the association between leukocytes and the MiOXSYS assay is entirely novel. We did not observe any difference in levels of sORP or DNA fragmentation in samples with or without leukocytospermia, which was unexpected, especially as leukocytes are a significant source of ROS.

The correlation between the two methods for measuring OS is not strong and this is not surprising since the assays measure different aspects of OS. Chemiluminescence measurement of OS specifically detects ROS [51,52]. In contrast, measurement of redox potential generates a single measurement from the culmination of the plethora of oxidation-reduction reactions occurring within a biological system [53]. Given that the two methods assess OS from different perspectives, it is important to establish that they are both relevant biomarkers of sperm pathology in order for them to be useful as diagnostic tests for infertility.

There is a moderate correlation between ROS and DFI and it is highly significant, while the correlation between sORP and DFI is not significant. Only one previous study has examined the relationship between sORP and DNA fragmentation [45], but DNA damage was measured by sperm chromatin dispersion rather than SCSA as in this study. They showed that while sORP and sperm DFI were not correlated in fertile men, there was a significant correlation in infertile men [45]. Since the fertility status of the men in this study was unknown, it is possible that the samples assessed for sORP contained a higher proportion of men who were fertile. The relationship between OS and DFI becomes considerably more apparent when samples are grouped according to whether they are considered to be in oxidative balance or OS. Under these circumstances, there is a significant increase in sperm DNA damage in semen samples that exhibit OS, whether OS is measured by ROS or sORP, although this is particularly significant when OS is measured by sORP. There is a wealth of evidence demonstrating an association between sperm DNA damage and OS assessed by chemiluminescence [40,42–44,52], but this is the first study to report an increase in sperm DNA damage in samples exposed to OS as determined by the MiOXSYS assay.

Oxidative stress is well known to manifest its effects on semen parameters [28,38–41,46], as well as playing a major role in sperm DNA damage [8,21,22,27,30,40,42–44,54]. In this study, while we have shown that both sORP and ROS are significantly negatively correlated with semen parameters, the correlation is approximately twice as strong for all parameters, with the exception of vitality, when OS is measured by sORP. Sperm DNA fragmentation is also significantly negatively correlated with all semen parameters, particularly with motility and vitality, in agreement with previous findings [49,55,56]. Interestingly, sperm DNA fragmentation shows a much stronger negative correlation than OS with sperm motility and vitality, indicating an alternative source for DNA damage that does not involve OS. Indeed, high DFI levels in some samples are not accompanied by elevations in sORP or ROS. Although sperm DNA damage and subsequent loss of vitality is a major consequence of OS [54], oxidative damage is only one of several etiologies that are responsible for DNA fragmentation, including abnormal caspase activity leading to abortive apoptosis, incomplete protamination and chromatin packaging abnormalities and anomalies in endonuclease and topoisomerase II activity [6,8,11].

The association of OS and sperm DNA damage with semen parameters is further highlighted in this study by the higher levels of these biomarkers seen in semen from patients with abnormal semen parameters compared to those with normal semen parameters. A significant reduction in semen parameters and a significant increase in seminal ROS have previously been demonstrated in infertile men [28]. In addition, sORP was elevated in men with abnormal semen parameters [57], and a significant correlation between morphology and sORP has been demonstrated in infertile men [45]. We have shown that samples with abnormal semen parameters are more likely to have elevated sORP and ROS, with increased DFI and HDS. However, the relationship between sORP, ROS and DNA damage is more complex and may be dependent upon the source of oxidants. The results presented in this study corroborate previous findings demonstrating excessive production of ROS by PMN [12,14,15,33,41,58], but on the contrary, sORP is not elevated in samples with leukocytospermia, and there is no evidence for any significant increase in sperm DNA fragmentation. On the other hand, measurement of sperm DNA fragmentation is not a true measurement of DNA oxidation as it assesses DNA strand breaks only, and oxidation of sperm DNA with ensuing pathology may occur in the absence of any measurable levels of sperm DNA fragmentation. Furthermore, the presence of leukocytes does not appear to cause any damage to sperm parameters in this study. The role of leukocytes in male infertility remains controversial. Some studies reveal no correlation between seminal PMN and semen parameters [59] or sperm DNA damage [14,58,59] assessed by either TUNEL or 8–OHDG, while two studies demonstrated a significant positive correlation between PMN and DNA damage [12,60] assessed by TUNEL and SCSA respectively, and two studies showed negative correlations between PMN and semen parameters [12,58]. One explanation for this could be that there is a temporal lag between release of ROS and the time at which physiological effects in sperm are manifested. Since the origin of seminal leukocytes is primarily from the male accessory glands [61], exposure of spermatozoa to ROS produced by leukocytes would only occur at ejaculation, with insufficient time to have any significant effect on semen quality. However, this does not explain the low levels of sORP in samples with leukocytospermia, even though PMN are producing exceptionally high levels of ROS.

Alternatively, it could be argued that the detrimental effects of OS on sperm are due to the location of ROS production. ROS are produced exogenously by PMN within the seminal fluid milieu, which is normally enriched in antioxidants [62]. Hence the effects of the oxidants are mitigated before they come into contact with the sperm. Alternatively, OS produced intrinsically by sperm themselves are more likely to cause DNA damage as spermatozoa contain negligible antioxidants making the internal components highly susceptible to oxidative damage [19]. Taken together, these observations may serve to explain why seminal ROS levels are high and sORP levels are relatively low in the presence of PMN, since PMN are a source of exogenous ROS, and sORP measurement takes into account the levels of antioxidants as well as pro-oxidants in the system. It may also explain why the correlation between OS and semen parameters is much stronger when OS is measured by sORP, and why the increase in DNA damage in the group with OS is more significant when OS is measured using sORP.

Another marker of sperm genetic integrity is that of high DNA stainability (HDS), which measures sperm with abnormal protamination of chromatin where nuclear histones are retained. We observed a strong negative correlation between HDS and sperm morphology confirming previous observations that sperm head abnormalities can be caused by protamine deficiency, incomplete protamine sulfhydryl oxidation and chromatin condensation [63,64]. The origin of aberrant protamination in the sperm nucleus is unclear. Although we did not observe any correlation between HDS and OS, HDS levels were increased in samples with OS compared to those that did not exhibit OS, but the difference is only significant when OS is measured by ROS. Furthermore, unlike DFI, HDS is significantly higher in leukocytospermia consistent with previous studies confirming a role for extrinsic ROS in sperm nuclear chromatin compaction [65,66].

Overall, the stronger association of sORP with both DNA fragmentation and semen parameters lends support to the view that measurement of redox potential is a more powerful tool than chemiluminescence for determining the pathological oxidative state of the sperm. The severity of the pathological consequences of oxidative stress on sperm highlights the importance of measuring OS as the most influential marker of sperm function. However, more work is warranted to establish the proposed theoretical model of extrinsic and intrinsic ROS production and whether leukocytospermia contributes to sperm DNA fragmentation.

Author Contributions: Authors S.T.H. and J.W.A.R. were involved in the conception and design of the study and patient recruitment. A.M.V., J.S., A.P.K., A.D., F.G. and J.X. carried out the investigations and collected the data. Validations were performed by AV and JS. Data analysis was performed by A.P.K., J.S., A.P.K., and S.T.H. S.T.H. and A.P.K. supervised the project. S.T.H. wrote the manuscript. All authors have read and approved the final manuscript.

Funding: Sensors for MiOXSYS assays were generously provided by Aytu Bioscience Inc. No additional funding was required for this study.

Conflicts of Interest: The authors declare no conflicts of interest.

References

1. Darbandi, M.; Darbandi, S.; Agarwal, A.; Sengupta, P.; Durairajanayagam, D.; Henkel, R.; Sadeghi, M. Reactive Oxygen Species and Male Reproductive Hormones. *Reprod. Biol. Endocrinol.* **2018**, *16*, 87. [CrossRef] [PubMed]
2. Wagner, H.; Cheng, J.; Ko, E. Role of Reactive Oxygen Species in Male Infertility: An Updated Review of Literature. *Arab J. Urol.* **2017**, *16*, 35–43. [CrossRef] [PubMed]
3. Wright, C.; Milne, S.; Leeson, H. Sperm DNA Damage Caused by Oxidative Stress: Modifiable Clinical, Lifestyle and Nutritional Factors in Male Infertility. *Reproduct. BioMed. Online* **2014**, *28*, 684–703. [CrossRef] [PubMed]
4. Tremellen, K. Oxidative Stress and Male Infertility—A Clinical Perspective. *Hum. Reproduct. Update* **2008**, *14*, 243–258. [CrossRef] [PubMed]
5. Dada, R.; Bisht, S. Oxidative Stress Major Executioner in Disease Pathology Role in Sperm DNA Damage and Preventive Strategies. *Front. Biosci.* **2017**, *9*, 420–447. [CrossRef]
6. Aitken, R. Reactive Oxygen Species as Mediators of Sperm Capacitation and Pathological Damage. *Mol. Reproduct. Dev.* **2017**, *84*, 1039–1052. [CrossRef]
7. Du Plessis, S.; Agarwal, A.; Halabi, J.; Tvrda, E. Contemporary Evidence on The Physiological Role of Reactive Oxygen Species In Human Sperm Function. *J. Assist. Reproduct. Genet.* **2015**, *32*, 509–520. [CrossRef] [PubMed]
8. Aitken, R.; Curry, B. Redox Regulation of Human Sperm Function: From the Physiological Control of Sperm Capacitation to the Etiology of Infertility and DNA Damage in The Germ Line. *Antioxid. Redox Signal.* **2011**, *14*, 367–381. [CrossRef] [PubMed]
9. de Lamirande, E.; Jiang, H.; Zini, A.; Kodama, H.; Gagnon, C. Reactive Oxygen Species and Sperm Physiology. *Rev. Reproduct.* **1997**, *2*, 48–54. [CrossRef]
10. O'Flaherty, C.; Matsushita-Fournier, D. Reactive Oxygen Species and Protein Modifications in Spermatozoa. *Biol. Reproduct.* **2017**, *97*, 577–585. [CrossRef]
11. Sakkas, D.; Alvarez, J. Sperm DNA Fragmentation: Mechanisms of Origin, Impact on Reproductive Outcome, and Analysis. *Fertil. Steril.* **2010**, *93*, 1027–1036. [CrossRef] [PubMed]
12. Lobascio, A.; De Felici, M.; Anibaldi, M.; Greco, P.; Minasi, M.; Greco, E. Involvement of Seminal Leukocytes, Reactive Oxygen Species, and Sperm Mitochondrial Membrane Potential in the DNA Damage of the Human Spermatozoa. *Andrology* **2015**, *3*, 265–270. [CrossRef]
13. Aitken, R.; Baker, M. Oxidative Stress, Spermatozoa and Leukocytic Infiltration: Relationships Forged by the Opposing Forces of Microbial Invasion and the Search for Perfection. *J. Reproduct. Immunol.* **2013**, *100*, 11–19. [CrossRef]
14. Mupfiga, C.; Fisher, D.; Kruger, T.; Henkel, R. The Relationship Between Seminal Leukocytes, Oxidative Status in the Ejaculate, and Apoptotic Markers in Human Spermatozoa. *Syst. Biol. Reproduct. Med.* **2013**, *59*, 304–311. [CrossRef] [PubMed]
15. Henkel, R. Leukocytes and Oxidative Stress: Dilemma for Sperm Function and Male Fertility. *Asian J. Androl.* **2011**, *13*, 43–52. [CrossRef]
16. Koppers, A.; De Iuliis, G.; Finnie, J.; McLaughlin, E.; Aitken, R. Significance of Mitochondrial Reactive Oxygen Species in the Generation of Oxidative Stress in Spermatozoa. *J. Clin. Endocrinol. Metab.* **2008**, *93*, 3199–3207. [CrossRef] [PubMed]
17. Ford, W. Regulation of Sperm Function by Reactive Oxygen Species. *Hum. Reproduct. Update* **2004**, *10*, 387–399. [CrossRef] [PubMed]
18. Gomez, E.; Buckingham, D.; Brindle, J.; Lanzafame, F.; Irvine, D.; Aitken, R. Development of an Image Analysis System to Monitor the Retention of Residual Cytoplasm by Human Spermatozoa: Correlation With Biochemical Markers of the Cytoplasmic Space, Oxidative Stress, and Sperm Function. *J. Androl.* **1996**, *17*, 276–287.
19. Aitken, R.; Gibb, Z.; Baker, M.; Drevet, J.; Gharagozloo, P. Causes and Consequences of Oxidative Stress in Spermatozoa. *Reproduct. Fertil. Dev.* **2016**, *28*, 1–10. [CrossRef]
20. Moazamian, R.; Polhemus, A.; Connaughton, H.; Fraser, B.; Whiting, S.; Gharagozloo, P.; Aitken, R. Oxidative Stress and Human Spermatozoa: Diagnostic and Functional Significance of Aldehydes Generated as a Result of Lipid Peroxidation. *MHR Basic Sci. Reproduct. Med.* **2015**, *21*, 502–515. [CrossRef]

21. Muratori, M.; Tamburrino, L.; Marchiani, S.; Cambi, M.; Olivito, B.; Azzari, C.; Forti, G.; Baldi, E. Investigation on the Origin of Sperm DNA Fragmentation: Role of Apoptosis, Immaturity and Oxidative Stress. *Mol. Med.* **2015**, *21*, 109–122. [CrossRef] [PubMed]

22. Aitken, R.; Koppers, A. Apoptosis and DNA Damage in Human Spermatozoa. *Asian J. Androl.* **2010**, *13*, 36–42. [CrossRef] [PubMed]

23. Darbandi, M.; Darbandi, S.; Agarwal, A.; Baskaran, S.; Dutta, S.; Sengupta, P.; Khorram Khorshid, H.; Esteves, S.; Gilany, K.; Hedayati, M.; et al. Reactive Oxygen Species-Induced Alterations in H_{19}-Igf_2 Methylation Patterns, Seminal Plasma Metabolites, and Semen Quality. *J. Assist. Reproduct. Genet.* **2018**. [CrossRef] [PubMed]

24. Menezo, Y.; Silvestris, E.; Dale, B.; Elder, K. Oxidative Stress and Alterations in DNA Methylation: Two Sides of the Same Coin in Reproduction. *Reproduct. BioMed. Online* **2016**, *33*, 668–683. [CrossRef]

25. Tunc, O.; Tremellen, K. Oxidative DNA Damage Impairs Global Sperm DNA Methylation in Infertile Men. *J. Assist. Reproduct. Genet.* **2009**, *26*, 537–544. [CrossRef] [PubMed]

26. Santi, D.; Spaggiari, G.; Simoni, M. Sperm DNA Fragmentation Index as a Promising Predictive Tool for Male Infertility Diagnosis and Treatment Management—Meta-Analyses. *Reproduct. BioMed. Online* **2018**, *37*, 315–326. [CrossRef]

27. Khosravi, F.; Valojerdi, M.; Amanlou, M.; Karimian, L.; Abolhassani, F. Relationship of Seminal Reactive Nitrogen and Oxygen Species and Total Antioxidant Capacity with Sperm DNA Fragmentation in Infertile Couples with Normal and Abnormal Sperm Parameters. *Andrologia* **2014**, *46*, 17–23. [CrossRef] [PubMed]

28. Agarwal, A.; Sharma, R.; Sharma, R.; Assidi, M.; Abuzenadah, A.; Alshahrani, S.; Durairajanayagam, D.; Sabanegh, E. Characterizing Semen Parameters and Their Association with Reactive Oxygen Species in Infertile Men. *Reproduct. Biol. Endocrinol.* **2014**, *12*, 33–42. [CrossRef] [PubMed]

29. Zini, A.; Lamirande, E.; Gagnon, C. Reactive Oxygen Species in Semen of Infertile Patients: Levels of Superoxide Dismutase- And Catalase-Like Activities in Seminal Plasma and Spermatozoa. *Int. J. Androl.* **1993**, *16*, 183–188. [CrossRef]

30. Kamkar, N.; Ramezanali, F.; Sabbaghian, M. The Relationship Between Sperm DNA Fragmentation, Free Radicals and Antioxidant Capacity with Idiopathic Repeated Pregnancy Loss. *Reproduct. Biol.* **2018**, *18*, 330–335. [CrossRef]

31. Leach, M.; Aitken, R.; Sacks, G. Sperm DNA Fragmentation Abnormalities in Men from Couples with a History of Recurrent Miscarriage. *Aust. N. Z. J. Obstet. Gynaecol.* **2015**, *55*, 379–383. [CrossRef]

32. Evenson, D. Sperm Chromatin Structure Assay (SCSA®). In *Spermatogenesis. Methods in Molecular Biology (Methods and Protocols)*; Carrell, D., Aston, K., Eds.; Humana Press: Totowa, NJ, USA, 2013; pp. 147–164.

33. Homa, S.; Vessey, W.; Perez-Miranda, A.; Riyait, T.; Agarwal, A. Reactive Oxygen Species (ROS) In Human Semen: Determination of a Reference Range. *J. Assist. Reproduct. Genet.* **2015**, *32*, 757–764. [CrossRef]

34. Vessey, W.; Perez-Miranda, A.; Macfarquhar, R.; Agarwal, A.; Homa, S. Reactive Oxygen Species in Human Semen: Validation and Qualification of a Chemiluminescence Assay. *Fertil. Steril.* **2014**, *102*, 1576–1583. [CrossRef]

35. Saleh, R.; Agarwal, A. Oxidative Stress and Male Infertility: From Research Bench to Clinical Practice. *J. Androl.* **2002**, *23*, 737–752.

36. Agarwal, A.; Bui, A. Oxidation-Reduction Potential as a New Marker for Oxidative Stress: Correlation to Male Infertility. *Investig. Clin. Urol.* **2017**, *58*, 385–399. [CrossRef]

37. Bar-Or, D.; Bar-Or, R.; Rael, L.; Brody, E. Oxidative Stress in Severe Acute Illness. *Redox Biol.* **2015**, *4*, 340–345. [CrossRef]

38. Takeshima, T.; Yumura, Y.; Yasuda, K.; Sanjo, H.; Kuroda, S.; Yamanaka, H.; Iwasaki, A. Inverse Correlation Between Reactive Oxygen Species in Unwashed Semen and Sperm Motion Parameters as Measured by a Computer-Assisted Semen Analyzer. *Asian J. Androl.* **2017**, *19*, 350–354. [CrossRef]

39. Pasqualotto, F.; Sharma, R.; Nelson, D.; Thomas, A.; Agarwal, A. Relationship Between Oxidative Stress, Semen Characteristics, and Clinical Diagnosis in Men Undergoing Infertility Investigation. *Fertil. Steril.* **2000**, *73*, 459–464. [CrossRef]

40. Barroso, G.; Morshedi, M.; Oehninger, S. Analysis of DNA Fragmentation, Plasma Membrane Translocation of Phosphatidylserine and Oxidative Stress in Human Spermatozoa. *Hum. Reproduct.* **2000**, *15*, 1338–1344. [CrossRef]

41. Aitken, R.; Buckingham, D.; Brindle, J.; Gomez, E.; Baker, H.; Irvine, D. Andrology: Analysis of Sperm Movement in Relation to the Oxidative Stress Created by Leukocytes in Washed Sperm Preparations and Seminal Plasma. *Hum. Reproduct.* **1995**, *10*, 2061–2071. [CrossRef]

42. Khodair, H.; Omran, T. Evaluation of Reactive Oxygen Species (ROS) and DNA Integrity Assessment in Cases of Idiopathic Male Infertility. *Egypt. J. Dermatol. Venerol.* **2013**, *33*, 51–55. [CrossRef]

43. Saleh, R.; Agarwal, A.; Nada, E.; El-Tonsy, M.; Sharma, R.; Meyer, A.; Nelson, D.; Thomas, A. Negative Effects of Increased Sperm DNA Damage in Relation to Seminal Oxidative Stress in Men with Idiopathic and Male Factor Infertility. *Fertil. Steril.* **2003**, *79*, 1597–1605. [CrossRef]

44. Lopes, S.; Jurisicova, A.; Sun, J.; Casper, R. Reactive Oxygen Species: Potential Cause for DNA Fragmentation in Human Spermatozoa. *Hum. Reproduct.* **1998**, *13*, 896–900. [CrossRef]

45. Majzoub, A.; Arafa, M.; Mahdi, M.; Agarwal, A.; Al Said, S.; Al-Emadi, I.; El Ansari, W.; Alattar, A.; Al Rumaihi, K.; Elbardisi, H. Oxidation–Reduction Potential and Sperm DNA Fragmentation, and Their Associations with Sperm Morphological Anomalies Amongst Fertile and Infertile Men. *Arab J. Urol.* **2018**, *16*, 87–95. [CrossRef]

46. Arafa, M.; Agarwal, A.; Al Said, S.; Majzoub, A.; Sharma, R.; Bjugstad, K.; AlRumaihi, K.; Elbardisi, H. Semen Quality and Infertility Status Can Be Identified Through Measures of Oxidation-Reduction Potential. *Andrologia* **2017**, *50*, e12881. [CrossRef]

47. World Health Organization. *WHO Laboratory Manual for the Examination and Processing of Human Semen*, 5th ed.; WHO Press: Geneva, Switzerland, 2010.

48. Rael, L.; Bar-Or, R.; Kelly, M.; Carrick, M.; Bar-Or, D. Assessment of Oxidative Stress in Patients with an Isolated Traumatic Brain Injury Using Disposable Electrochemical Test Strips. *Electroanalysis* **2015**, *27*, 2567–2573. [CrossRef]

49. Stahl, P.; Cogan, C.; Mehta, A.; Bolyakov, A.; Paduch, D.; Goldstein, M. Concordance Among Sperm Deoxyribonucleic Acid Integrity Assays and Semen Parameters. *Fertil. Steril.* **2015**, *104*, 56–61. [CrossRef]

50. Agarwal, A.; Qiu, E.; Sharma, R. Laboratory Assessment of Oxidative Stress in Semen. *Arab J. Urol.* **2018**, *16*, 77–86. [CrossRef]

51. Wardman, P. Fluorescent and Luminescent Probes for Measurement of Oxidative and Nitrosative Species in Cells and Tissues: Progress, Pitfalls, and Prospects. *Free Radic. Biol. Med.* **2007**, *43*, 995–1022. [CrossRef]

52. Aitken, R.; Baker, M.; O'Bryan, M. Andrology Lab Corner*: Shedding Light on Chemiluminescence: The Application of Chemiluminescence in Diagnostic Andrology. *J. Androl.* **2004**, *25*, 455–465. [CrossRef]

53. Spanidis, Y.; Mpesios, A.; Stagos, D.; Goutzourelas, N.; Bar-Or, D.; Karapetsa, M.; Zakynthinos, E.; Spandidos, D.; Tsatsakis, A.; Leon, G.; et al. Assessment of the Redox Status in Patients with Metabolic Syndrome and Type 2 Diabetes Reveals Great Variations. *Exp. Ther. Med.* **2016**, *11*, 895–903. [CrossRef]

54. Aitken, R.; De Iuliis, G.; Finnie, J.; Hedges, A.; McLachlan, R. Analysis of the Relationships Between Oxidative Stress, DNA Damage and Sperm Vitality in a Patient Population: Development of Diagnostic Criteria. *Hum. Reproduct.* **2010**, *25*, 2415–2426. [CrossRef]

55. Smit, M.; Romijn, J.; Wildhagen, M.; Weber, R.; Dohle, G. Sperm Chromatin Structure is Associated with the Quality of Spermatogenesis in Infertile Patients. *Fertil. Steril.* **2010**, *94*, 1748–1752. [CrossRef]

56. Moskovtsev, S.; Willis, J.; White, J.; Mullen, J. Sperm DNA Damage: Correlation to Severity of Semen Abnormalities. *Urology* **2009**, *74*, 789–793. [CrossRef]

57. Agarwal, A.; Wang, S. Clinical Relevance of Oxidation-Reduction Potential in the Evaluation of Male Infertility. *Urology* **2017**, *104*, 84–89. [CrossRef]

58. Henkel, R.; Kierspel, E.; Stalf, T.; Mehnert, C.; Menkveld, R.; Tinneberg, H.; Schill, W.; Kruger, T. Effect of Reactive Oxygen Species Produced by Spermatozoa and Leukocytes on Sperm Functions in Non-Leukocytospermic Patients. *Fertil. Steril.* **2005**, *83*, 635–642. [CrossRef]

59. Micillo, A.; Vassallo, M.; Cordeschi, G.; D'Andrea, S.; Necozione, S.; Francavilla, F.; Francavilla, S.; Barbonetti, A. Semen Leukocytes and Oxidative-Dependent DNA Damage of Spermatozoa in Male Partners of Subfertile Couples With No Symptoms of Genital Tract Infection. *Andrology* **2016**, *4*, 808–815. [CrossRef]

60. Alvarez, J.G.; Sharma, R.K.; Ollero, M.; Saleh, R.A.; Lopez, M.C.; Thomas, A.J., Jr.; Evenson, D.P.; Agarwal, A. Increased DNA Damage in Sperm from Leukocytospermic Semen Samples as Determined by the Sperm Chromatin Structure Assay. *Fertil. Steril.* **2002**, *7*, 319–329. [CrossRef]

61. Simbini, T.; Umapathy, E.; Jacobus, E.; Tendaupenyu, G.; Mbizvo, M. Study on the Origin of Seminal Leucocytes Using Split Ejaculate Technique and the Effect of Leucocytospermia on Sperm Characteristics. *Urol. Int.* **1998**, *61*, 95–100. [CrossRef]

62. Halliwell, B.; Gutteridge, J. The Antioxidants of Human Extracellular Fluids. *Arch. Biochem. Biophys.* **1990**, *280*, 1–8. [CrossRef]

63. Ma, Y.; Xie, N.; Li, Y.; Zhang, B.; Xie, D.; Zhang, W.; Li, Q.; Yu, H.; Zhang, Q.; Ni, Y.; et al. Teratozoospermia with Amorphous Sperm Head Associate with Abnormal Chromatin Condensation in a Chinese Family. *Syst. Biol. Reproduct. Med.* **2018**, *19*, 1–10. [CrossRef] [PubMed]

64. Zini, A.; Phillips, S.; Courchesne, A.; Boman, J.; Baazeem, A.; Bissonnette, F.; Kadoch, I.; San Gabriel, M. Sperm Head Morphology is Related to High Deoxyribonucleic Acid Stainability Assessed by Sperm Chromatin Structure Assay. *Fertil. Steril.* **2009**, *91*, 2495–2500. [CrossRef] [PubMed]

65. Aitken, R.J.; De Iuliis, G.N. On the Possible Origins of DNA Damage in Human Spermatozoa. *Mol. Hum. Reproduct.* **2010**, *16*, 3–13. [CrossRef]

66. De Iuliis, G.N.; Thomson, L.K.; Mitchell, L.A.; Finnie, J.M.; Koppers, A.J.; Hedges, A.; Nixon, B.; Aitken, R.J. DNA Damage in Human Spermatozoa is Highly Correlated with the Efficiency of Chromatin Remodeling and the Formation of 8-hydroxy-20-deoxyguanosine, a Marker of Oxidative Stress. *Biol. Reproduct.* **2009**, *81*, 517–524. [CrossRef]

Article

Human Sperm Chromosomes: To Form Hairpin-Loops, Or Not to Form Hairpin-Loops, That Is the Question

Dimitrios Ioannou [1,†] **and Helen G. Tempest** [1,2,*]

[1] Department of Human and Molecular Genetics, Herbert Wertheim College of Medicine, Florida International University, Miami, FL 33199, USA
[2] Biomolecular Sciences Institute, Florida International University, Miami, FL 33199, USA
* Correspondence: htempest@fiu.edu
† Current address: IVF Florida Reproductive Associates, Margate, FL, 33063, USA.

Received: 27 May 2019; Accepted: 28 June 2019; Published: 3 July 2019

Abstract: Background: Genomes are non-randomly organized within the interphase nucleus; and spermatozoa are proposed to have a unique hairpin-loop configuration, which has been hypothesized to be critical for the ordered exodus of the paternal genome following fertilization. Recent studies suggest that the hairpin-loop model of sperm chromatin organization is more segmentally organized. The purpose of this study is to examine the 3D organization and hairpin-loop configurations of chromosomes in human spermatozoa. Methods: Three-color sperm-fluorescence in-situ hybridization was utilized against the centromeres, and chromosome p- and q-arms of eight chromosomes from five normozoospermic donors. Wide-field fluorescence microscopy and 3D modelling established the radial organization and hairpin-loop chromosome configurations in spermatozoa. Results: All chromosomes possessed reproducible non-random radial organization ($p < 0.05$) and formed discrete hairpin-loop configurations. However, chromosomes preferentially formed narrow or wide hairpin-loops. We did not find evidence to support the existence of a centralized chromocenter(s) with centromeres being more peripherally localized than one or both of their respective chromosome arms. Conclusion: This provides further evidence to support a more segmental organization of chromatin in the human sperm nucleus. This may be of significance for fertilization and early embryogenesis as specific genomic regions are likely to be exposed, remodeled, and activated first, following fertilization.

Keywords: nuclear organization; chromatin; spermatozoa; chromosomes; chromosome territories; centromeres

---

1. Introduction

1.1. Organization of the Human Genome

The human genome consists of 23 pairs of chromosomes, each chromosome pair is in essence a unique linear sequence of DNA nucleotides that differs in length, with chromosome 1 being the longest (~250 Mbp), and chromosome 21 being the shortest (~48 Mbp). The linear sequence of our genome is housed in the cell nucleus (~10 micrometers). If DNA were to be unraveled, and disassociated from various proteins it would reach almost 2 meters in length. Somehow, the considerable amount of information contained in DNA must be efficiently packaged so that the genome can not only fit within the nucleus, but also in a way that facilitates a myriad of required normal cellular functions (e.g., DNA transcription, DNA replication, DNA damage recognition and repair etc.) [1]. So, how is our genome packaged in a functional manner? If asked to describe the packaging of DNA many would describe what is most often portrayed in textbooks. This classical view often depicts each linear sequence of

DNA being efficiently packaged by wrapping itself around histones to form a nucleosome, which is then packaged to form a solenoid, which is further packaged into various loops and domains. During the metaphase stage of cell division, DNA is further packaged and maximally condensed into highly organized structures, known as chromosomes. Frequently, (particularly for cytogeneticists), when we think about these linear strands of DNA the mental picture conjured up is that of a chromosome with its classical features including the telomeres, centromere, and the short (p) and long (q) arms. However, it is important to recognize that the metaphase chromosome only exists for a very short period of the cell cycle. So what happens to the packaging and organization of these linear pieces of DNA during the remainder of the cell cycle when it is more decondensed?

The genome even during interphase, when the DNA is at its most decondensed, remains highly organized with chromosomes forming distinct chromosome territories (CTs). The existence of CTs is not a new concept and was in fact, first proposed over 110 years ago by Theodore Boveri following his studies of blastomeres in horse roundworms [2]. However, the concept of CTs was largely forgotten until the work of Thomas Cremer and others in the 1970s and 1980s (reviewed in [3–5]). The existence and organization of CTs in the interphase nucleus has been experimentally established for many decades and widely accepted in scientific literature [6]. Nevertheless, the majority of texts still erroneously omit the existence of CTs when depicting the structure and packaging of DNA. Many studies have utilized fluorescence in-situ hybridization (FISH) to study the spatial organization of CTs. The majority of these studies have demonstrated that CTs exhibit distinct radial patterns of nuclear organization in somatic cells, which has been observed in two- and three-dimensions (2D and 3D). Of note, somatic chromatin organization has been shown to be cell-type specific [5], evolutionarily conserved [7], and reproducible between individuals [8–11]. Distinct radial patterns of CT organization have been reported for the majority of CTs in diverse cell-types and species. Radial CT organization typically follows the gene density or chromosome size model, in which CTs are preferentially radially organized based on either of these chromosome characteristics [12]. The unique organization of CTs has been postulated to serve as an additional layer of epigenetic regulation of the genome that likely plays an important role in normal cellular functions [13]. Largely, the radial organization of the majority of CTs in diverse cell-types and species typically follows one of the two models previously described. However, there is one notable exception, spermatozoa, which seemingly displays a unique model of organization, compared to other cell types. Given that spermatozoa are ellipsoid in shape and contain a flagellum, some organization studies have assessed radial and/or longitudinal organization of chromosomes or targeted loci [10,11,14–28]. The vast majority of these studies have reported non-random chromatin organization in sperm from various species, with the exception of chicken [26,27]. The majority of studies that have assessed the radial organization of chromosomes in sperm, suggest that the organization tends to follow the gene density model [11,16,17,22]. However, a distinct pattern of organization in sperm was proposed by the Zalensky group in the 1990s—the hairpin-loop model. This model described chromosome centromeres clustering in the center of the nucleus to form a single or multiple chromocenters, and the p- and q-arms of each chromosome lying in parallel with each other, or intermingled as they stretched from the center of the nucleus to the nuclear periphery where the telomeres formed dimers and tetramers (Figure 1A) [25,29–32]. This unique CT configuration was originally described by the Zalensky group as the hairpin-loop model, due to its resemblance of a hairpin-loop structure. The description of the hairpin-loop model led to this term being adopted and routinely utilized in the field to describe sperm chromatin organization. If the analogy of a bicycle wheel is used the centromeres would form the hub, the chromosome arms the spokes, and the telomeres the wheel rim. However, few studies have revisited this model of sperm chromatin organization since it was proposed over 20 years ago. Recently, we published a study that reexamined this model of organization utilizing both 2D and 3D approaches to assess the organization of centromeres and telomeres in normozoospermic males. The study findings led us to conclude that the original hairpin-loop model needed to be further refined. We provided interindividual reproducible evidence of multiple chromocenters that were not localized centrally and telomeres that were not restricted to the nuclear periphery as previously suggested, but rather centromeres and telomeres were localized throughout the nucleus. Thus, we have suggested that sperm chromatin organization is perhaps less like the bicycle wheel (Figure 1A) but rather more segmentally organized in the nucleus, with hairpin-loop structures forming in different directions and orientations in the nucleus (Figure 1B) [6,10].

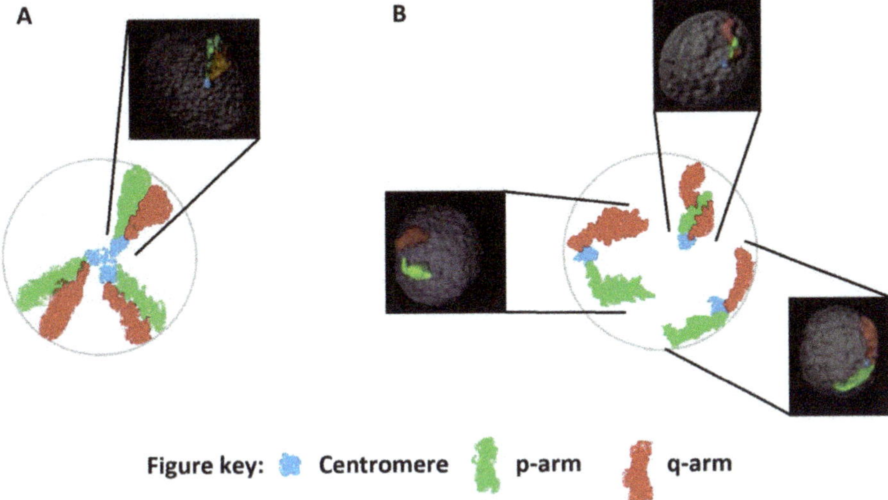

Figure 1. 3D rendered models and schematic diagrams illustrating two different models of chromosome organization and hairpin-loop configurations in human sperm. Schematic cross-section of a sperm nucleus, with magnified boxes that display 3D rendered models following sperm-FISH of the organization of centromeres (aqua), p-arms (green) and q-arms (red) in human sperm. Panel A: Displays the original chromocenter hairpin-loop model of sperm chromosome organization, whereby centromeres cluster in the nuclear interior of the sperm nucleus, forming one or more chromocenters, with the p- and q-arms stretching out toward the nuclear periphery. In the current study of 1240 cells, we rarely observed this type of chromosome configuration. This, alongside previously published data [10], led us to propose a refined hairpin-loop model of organization that is more segmentally arranged. Panel B: Segmental model of chromosome organization in human sperm. Here, we observe centromeres distributed throughout the nucleus and the chromosome arms forming narrow and wide hairpin-loops as well as displaying different orientations within the sperm nucleus. Additionally, in contrast to one study [29], we rarely observed chromosome arms to lie in parallel to one another leading to intermingling or coiling of the chromosome arms; with the majority of cells examined displaying discrete non-overlapping territories for the chromosome arms.

1.2. Why Do Sperm Exhibit A Different Organization to Somatic Cells?

The sperm cell is a highly specialized cell, whose main function is to transport safely the paternal genome during its long, arduous journey to the oocyte. The process of spermatogenesis is a unique process whereby spermatogonial stem cells produce diploid spermatogonia which ultimately undergo meiosis and differentiate to form four haploid spermatozoa. During spermiogenesis, some of the most dramatic transformations in cell biology take place. Haploid round spermatids undergo substantial remodeling and repackaging of the genome into one of the smallest cells; a streamlined spermatozoon, which also contains an acrosome cap, a mid-piece enriched with mitochondria and a flagellum [33]. Despite the efficient packaging of the genome in somatic cells, spermatozoa have evolved an even more effective mechanism through which to condense and package its genome. In humans, the majority (85–95%) of DNA associated histones are initially replaced with transition proteins and finally with protamines [34], with the remaining 5–15% of the genome retaining histone packaging [35]. The largely protamine packaged sperm genome has been proposed to serve several critical functions. These may include: (i) enabling a smaller cell volume and facilitating the hydrodynamic shape required to fulfill its function; (ii) protecting the paternal genome from DNA damage, as sperm lack DNA repair mechanisms; and (iii) deprogramming and inactivating the genome resulting in a relatively inert "silent" vessel [34,36]. Given the unique features of the sperm cell, it is perhaps not surprising that CTs display a unique spatial organization when compared to somatic cells.

1.3. What Could Be the Functional Consequence of Chromatin Organization in Spermatozoa?

Increasingly, evidence suggests that the sperm cell may not in fact be a "silent vessel"; but rather the sperm cell delivers an epigenetically primed genome to the maternal oocyte. Elegantly designed studies have demonstrated that histone bound regions of the paternal genome are more decondensed and are preferentially enriched in specific genes and gene promoters (including imprinted genes, developmentally important signaling proteins and transcription factors etc.) [20,37–40]. Furthermore, DNA hypomethylation of developmentally important gene families in the sperm cell may permit early transcriptional activation during early embryogenesis [34,38]. Recent high resolution molecular studies in mice have demonstrated that the transcriptional repressor CTCF binding motif is preferentially associated with the more condensed protamine packaged regions of the mouse genome [41]. Hi-C studies evaluating 3D interactions in murine gametes and early embryogenesis have identified distinct patterns between male and female gametes and embryos, with sperm containing more associations and long-range interactions than oocytes [42,43]. The unique, hierarchical CT organization throughout spermatogenesis is likely to be important for the creation of, and normal function of spermatozoa. However, many have proposed that CT organization in sperm is an integral aspect of epigenetic mechanisms required for early embryogenesis [6,13,44,45]. During fertilization, the epigenetically primed paternal genome is hypothesized to be gradually exposed and remodeled by the oocyte [23,31]. Thus, many have also proposed that sperm CT organization is an additional level of epigenetic programming that may play a crucial role in the formation of the male pronucleus and early embryonic development [6,18,29,31]. These findings suggest that we need to revisit the concept of spermatozoa being "silent vessels", rather spermatozoa carry an epigenetically primed paternal genome, which not only delivers critical information to the oocyte, but may be poised to activate critical genes for early embryogenesis following fertilization [41]. Furthermore, the CT organization most likely serves as another critical aspect of the epigenetically poised sperm nucleus. CT organization likely functions as a unique mechanism to deliver, unpackage, remodel, and transfer the genetic and epigenetic information from the paternal genome to the oocyte, perhaps in a highly ordered fashion [46]. Perhaps specific chromosomes or chromosomal regions need to be delivered sequentially to the maternal ooplasm to respond to oocyte signals to begin the formation of the male pronucleus [23]. Therefore, CT organization in the sperm nucleus may play a more important role in early embryonic divisions than currently perceived. Given the potential functional significance and unique CT organization in spermatozoa, we wanted to extend our initial 3D study of the organization of centromeres and telomeres to evaluate centromeres and the p- and q-arms of multiple chromosomes to establish whether we could confirm the existence of chromosome hairpin-loop formations. This was of particular interest given our newly proposed more segmentally organized sperm nucleus, which was based on centromere and telomere organization, and did not assess chromosome arm configuration and organization [6,10].

2. Materials and Methods

2.1. Patient Cohort and Semen Analysis

This study was conducted in accordance with the Declaration of Helsinki, and the protocol was approved by the Ethics Committee of Florida International University (FIU-IRB-13-0044). All methods were carried out in accordance with the approved guidelines. Cryopreserved semen samples were purchased from the Xytex Cryo International sperm bank. Five sperm donors of proven fertility were included in this study. Semen samples were collected via masturbation; and were classified as normozoospermic based on the World Health Organization criteria [47] semen parameter guidelines.

2.2. Semen Sample Preparation

Semen samples were prepared for FISH by removing the seminal fluid as described in detail previously [11]. In brief, cryopreserved semen samples were thawed at room temperature and washed with sperm wash buffer (10 mM NaCl, 10 mM Tris, pH 7.0) followed by centrifugation for 7 min at 504 g. Subsequently, the supernatant was removed and the pellet was resuspended with fresh sperm wash buffer and centrifuged as noted previously. Semen samples underwent a further 3–5 cycles of removal of the supernatant, addition of fresh wash buffer and centrifugation depending on the pellet size. Following the last sperm wash the supernatant was removed and the sample was then fixed drop-wise using 3:1 methanol:acetic acid to a final volume of 5 mL, the sample was then centrifuged as previously described and the fixation steps were repeated depending on the pellet size 3-5 times. Following fixation washes the pellet was resuspended in 1–1.5 mL of fixative depending on the pellet size and 1–3 µl of the sample was spread onto a glass slide. Slides were evaluated under a differential interference contrast microscope (Olympus BX53) for optimal cell density prior to initiation of FISH.

2.3. Sperm-FISH

The same FISH approach as described previously was applied in this study [10]. In brief, spermatozoa were spread onto glass microscope slides at the optimum density to minimize overlapping nuclei, cells then underwent a mild formaldehyde fixation to maintain the 3D structure of the nuclei as much as possible. This fixation step was followed by mild decondensation of sperm nuclei to facilitate FISH probe access in the densely protamine packaged sperm nuclei. This was achieved through a 20 min incubation in decondensation buffer (10 mM DTT (Sigma Aldrich, St Louis, MO, USA), 10 mM Tris solution, pH 8.0) in the dark. Following decondensation slides were rinsed in 2 × saline sodium citrate (SSC; Fisher Scientific, Pittsburgh, PA, USA), before dehydrating through an ethanol series (70–80–100%). Three-color FISH was performed for eight chromosomes, chromosomes: 2, 3, 6, 8, 10, 12, 16, and 18. FISH probe targets for each chromosome included the chromosome short (p) and long (q) arms, and satellite enumeration probes against the centromeres (Figure 2A). These specific chromosomes were chosen to evaluate a range of chromosome sizes and gene density; and to include both metacentric and submetacentric chromosomes, whilst excluding acrocentric chromosomes, which lack p-arms. Critically these specific chromosomes were also selected as centromere probes for these chromosomes do not cross hybridize with other centromeres. For each of the investigated chromosomes, arm paints were labelled in green (p-arms) and orange (q-arms) fluorochromes (MetaSystems, Boston, MA, USA) and centromeres were labelled with aqua fluorochromes (Kreatech, distributed by Leica Biosystems Buffalo Grove, IL, USA). FISH probes were utilized as per the manufacturer's guidelines. A 1:1:1 ratio mix of all 3 probes was denatured at 73 °C for 10 min and subsequently co-denatured with sperm cells for 90 s in a Thermobrite® Statspin (Abbott Molecular, Illinois, IL, USA) followed by hybridization for a minimum of 16 h at 37 °C. Post hybridization washes were carried out as per the manufacturers guidelines with the addition of a ddH$_2$O rinse for 1 min and an ethanol series (70–80–100%). Following these washes, slides were air dried, and subsequently mounted with 4′,6-diamidino-2-phenylindole (DAPI) antifade mounting medium (Vector Labs, Burlingame, CA, USA) under a 24 × 55 mm coverslip.

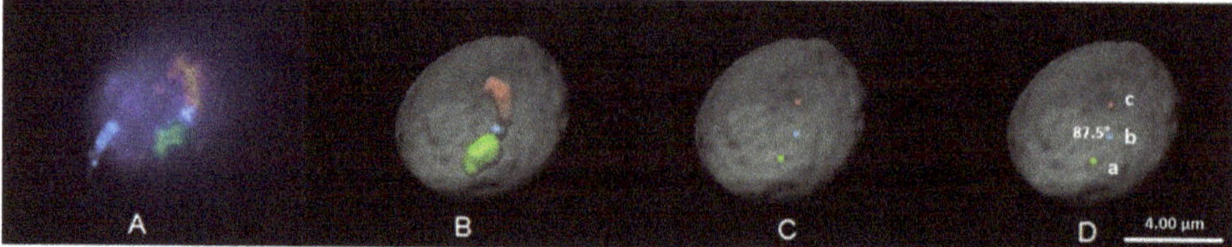

Figure 2. 3D FISH imaging and model rendering demonstrating the hairpin-loop configuration for chromosome 2 in a human sperm nucleus. FISH probes utilized corresponded to the centromere (aqua), p-arm (green), q-arm (red) of chromosome 2. The nucleus is counterstained with DAPI (blue: A) and is pseudo-colored (gray: B, C, and D) in 3D rendered models. (**A**) Depicts the raw FISH image following deconvolution (note: aqua signal at position 8 o'clock is background fluorescence from the sperm tail and is removed from 3D rendered models [panels B, C, and D], note the sperm tail was only visible in a small proportion of cells precluding assessment of longitudinal positioning). (**B**) Provides the 3D model reconstruction from the raw deconvolved image (A). (**C**) Depicts the geometric center for each of the three FISH probe targets as determined by the Imaris software. (**D**) Illustrates how the angle of the hairpin-loop configuration is calculated by measuring the angle formed between the p-arm (point a), and the q-arm (point c) through the centromere (point b), which in this cell is 87.5°. It is important to note that the angle created between the chromosome arms is determined in 3D models. In this figure the 3D data is reduced to 2D, which can give the impression that the geometric centers or FISH probes may lie in the same focal plane or z-stack, which is not the case rather this image has compressed the data from 92 sections taken at 0.2 µm intervals, into a single plane. Thus, the angle cannot be accurately assessed in the 2D image and the 3D model has to be utilized to account for differences in FISH probe localizations in different 3D focal planes (X, Y, Z). Scale bar is 4 µm.

2.4. 3D FISH Image Acquisition, Image Rendering and Analysis

The same methodology as previously reported was utilized to capture and render images in 3D [10]. Cells were imaged utilizing the DeltaVision high-resolution widefield fluorescence microscope (GE Healthcare Life Sciences, Pittsburgh, PA, USA) system; consisting of an Olympus IX71 inverted microscope with 60X, 1.4 NA oil-immersion lens and a photometric CCD. All images were taken with a Z step size of 0.2 µm (92 optical sections), saved as 3D stacks and subjected to constrained iterative deconvolution using the same standard settings (DeltaVision–SoftWoRx -V 5.5; GE Healthcare Life Sciences, Pittsburgh, PA, USA) (Figure 2A). A minimum of 30 images per subject, per probe set were acquired using TRITC (594 nm), FITC (523 nm), CFP (480 nm) filters. 3D stacks from SoftWoRx were reconstructed, rendered in 3D and analyzed utilizing Imaris software (V.7.6.3 Bitplane–Zurich, Switzerland) by converting images to 32-bit float images. The nuclear periphery was established using the DAPI counterstain and was rendered by creating a surface in Imaris. Similarly, the targeted loci (centromeres, p- and q-arms) were established based on fluorescence intensity of the FISH probes and translation of the pixel intensity resulted in a rendered surface (Figure 2B). To establish the nuclear periphery and model the FISH probes, an iso-surface was created to visualize the object in 3D space, whilst allowing verification of the accuracy directly against the original raw image. The creation of the nuclear surface was defined by setting an intensity threshold to select the voxels that were considered part of the reconstructed iso-surface. Minimal manual user intervention was required to establish the threshold, to best reflect the raw data set, and remove any background fluorescence. Voxel selection was further enhanced by applying a Gaussian filter prior to selection, to remove the noise not attributed to the labelled cell. This smoothing step adjusts for the limits of resolution of the acquisition system, and quality of the tissue labelling. This step ensures that the voxels from "out of focus light", which appear as a background blur in the Z-plane, were not included as a part of the surface structure ensuring a more accurate representation of the sperm nucleus. The reconstructed DAPI surface was overlaid on the raw image to ensure the created surfaces fitted the raw data in all three axes (X, Y, and Z). To measure the radial nuclear localization of target loci within the nucleus, the DAPI surface object was utilized to denote the nuclear periphery, this was used as a region of interest to isolate the other fluorescent channels (FISH signals) within the nucleus. The masking process generated a new channel based on the voxels that were located inside of the 3D volume of nuclear periphery for each individual fluorochrome. This new channel was rendered without interference from other voxels in the dataset, creating a new surface segmentation of structures within the nucleus. The Imaris Distance Transform (DT) tool utilizes a 3D quasi-Euclidean distance transformation from the geometric center (Figure 2C) of each rendered FISH signal in the data set to the binary mask of the DAPI border of the surface (nuclear periphery). The Imaris DT tool calculates the shortest distance in 3D space between each data point (FISH signals) and the DAPI nuclear periphery (surface border) with minimal user intervention, following the contours of the nucleus [10,48–50]. The Imaris DT tool is shape invariant and does not impose any assumptions on the cell shape or size [51]. To facilitate comparisons across experiments, it is desirable to have a measure that is both scale, and shape invariant; thus Imaris DT measurements were normalized against the widest "radial" diameter of individual sperm nuclei to account for differences between individual nuclei. The radius of each nuclei was divided by three to create three distinct nuclear regions (interior, intermediate, or peripheral). The radial organization of each FISH probe was assessed by measuring the distance of the geometric center of each loci to the nearest nuclear periphery as measured by the Imaris DT tool. The position of the geometric center of the FISH probe was then assigned to one of the three radial regions (interior, intermediate, or periphery). Hairpin-loop chromosome configurations were established by identifying the geometric center (Figure 2C) of each targeted loci (p-arm, q-arm, and the centromere) and determining the angle created between the p- and q-arms through the centromere (Figure 2D). Based on the angle created by the p- and q-arms, hairpin-loop configurations were arbitrarily stratified into two categories, those forming angles less than or equal to 40°, or greater than 40° to evaluate whether chromosomes had a tendency to form narrower or wider hairpin-loop configurations. The unique morphology of the sperm cell often facilitates assessment of longitudinal organization. Unfortunately, in this study the levels of background fluorescence were low, thus, rarely was the sperm tail visible (< 10% of analyzed cells) (Figure 2A), which precluded robust assessment of the longitudinal organization of the targeted loci. Thus, longitudinal organization or assessment of hairpin-loop configurations in relation to the sperm tail was not possible in the current study.

2.5. Statistical Analysis

The Chi-squared goodness of fit ($\chi2$) was utilized to evaluate if the radial organization of each target of interest differed from random, a *p*-value of < 0.05 suggested a non-random distribution. In essence, if the target loci were equally distributed across the three radial segments of the nucleus it suggested a random distribution, whereas preferential distribution in one or more segments as determined by the Chi-squared goodness of fit test indicated non-random organization of the target loci.

3. Results

In this study, we examined the 3D radial organization and chromosome configurations of eight different chromosomes in five normozoospermic subjects. A total of 1240 sperm were analyzed in this study, with a minimum of 30 nuclei evaluate per subject, per target loci (centromeres, p- and q-arms). Decondensation of nuclei should be avoided if at all possible when assessing the nuclear organization of chromatin. However, due to the unique protamine packaging in sperm, it is unfortunately a prerequisite for sperm-FISH. As published previously [10], we optimized the decondensation conditions to ensure efficient FISH hybridization (>95%) and minimal reproducible swelling between experiments and samples. A formaldehyde fixation step was performed prior to decondensation to maintain the 3D nuclear structure as much as possible. The decondensation parameters utilized in this study, resulted in a mild reproducible decondensation resulting in an average 2 fold increase in DAPI volume in decondensed sperm nuclei versus native sperm nuclei. Additionally, the increased nuclear volume in this study is comparable with, or lower than those previously reported in the literature [14,29].

3.1. Radial Organization of Chromosome Centromeres, p- and q-Arms in Sperm Nuclei

The radial organization of the geometric center of the three target loci (centromeres, p- and q-arms) for each of the eight investigated chromosomes (2, 3, 6, 8, 10, 12, 16, and 18) was evaluated by measuring the micrometer distance from the geometric center of each target to the nearest nuclear edge (Figure 3). The radius of the nucleus was utilized to normalize the distance measured for the geometric center of each probe target. The radius of each individual nucleus was divided into three regions (peripheral, intermediate, and interior) and the geometric center for each target loci was assigned to one of these three regions based on the distance from the nuclear periphery. All investigated chromosome targets were found to be non-randomly organized ($p < 0.05$) in the sperm nuclei from all five subjects with the exception of chromosome 12 centromere, which was found to be randomly organized ($p > 0.05$) (Figure 3C). Assessing the radial distribution of the three target loci for each chromosome in the three nuclear domains (interior, intermediate and periphery) several patterns emerge. The localization of the q-arms in each of the three nuclear regions is similar for all investigated chromosomes, ranging from ~17–24% found in the interior region, ~51–61% in the intermediate region, and ~20–25% in the nuclear periphery (Figure 3A). For the p-arms, a similar pattern of radial organization is observed for the majority of investigated chromosomes, ranging from ~16-31% found in the interior region, ~47–59% in the intermediate region, and ~19–29% in the nuclear periphery. However, the p-arms of chromosome 10 and 16 were localized more (~31%, chromosome 10) or less frequently (~16%, chromosome 16) in the nuclear interior in comparison to the other p-arms (Figure 3B). The radial positioning of the centromeres of the tested chromosomes appears much more variable when compared to the p- and q-arms, ranging from ~15–42% in the interior, ~34–61% in the intermediate region, and 18–29% in the nuclear periphery (Figure 3C). The radial position of centromeres 2, 3, 6, 8, 12, and 16 account for most of the variation observed particularly in the interior and intermediate regions of the nucleus.

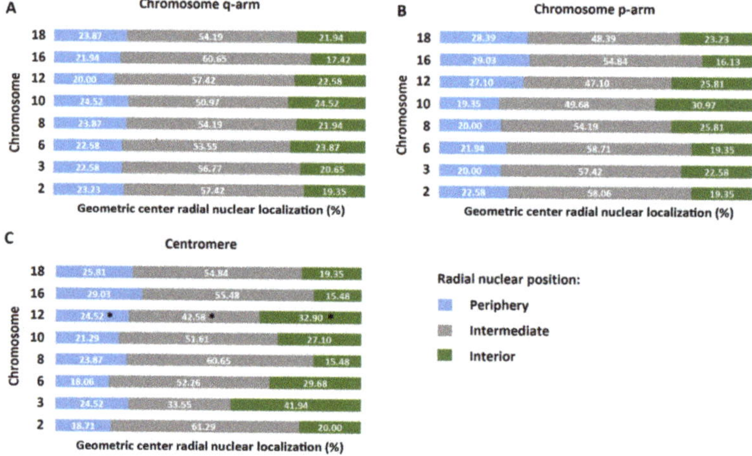

Figure 3. Mean 3D radial positioning of the centromeres, p- and q-arms of eight different chromosomes in human sperm from five normozoospermic subjects demonstrates relative consistent radial organization for the p- and q-arms, with more variation in organization for the centromeres. A minimum of 30 nuclei were examined per FISH probe target, per patient. The data displayed is based on the mean nuclear distribution of the geometric center of each probe targeted in a minimum of a 150 cells in the five patients enrolled in the study. All chromosome centromeres, p- and q-arms were non-randomly distributed in sperm nuclei ($p \leq 0.05$; χ^2 goodness-of-fit) with the exception of the chromosome 12 centromere which was randomly organized ($p \geq 0.05$) *. The nuclear positions of the geometric centers of the q-arms, p-arms and centromeres for each of the eight investigated chromosomes is shown in panels **A**, **B** and **C**, respectively.

To establish whether we could provide evidence of centralized chromocenters in human sperm nuclei, we utilized the micrometer measurements from the geometric center of each target loci to the nearest nuclear edge (Table 1). None of the investigated chromosomes strongly supported the concept of a centralized chromocenter, whereby one would expect centromeres to be more centrally localized than their respective chromosome arms (Table 1). When analyzing the raw data from the 155 cells studied in the five subjects, reproducible radial patterns of organization emerged for each the investigated loci. The data from the 1240 individual sperm cells analyzed revealed that centromeres were preferentially more distally localized in the sperm nucleus when compared to their respective chromosome arms, with 30.5% of centromeres being more distally located than either the p- or q-arm, and 37.8% being more distally located than both p- and q-arms (data not shown). Thus, the centromeres were found to be more peripherally localized in the sperm nucleus than at least one chromosome arm more than two-thirds of the time. Looking at the mean data presented in Table 1 from the five subjects, five out of the eight investigated centromeres (chromosomes 3, 6, 12, 16, and 18) were more peripherally localized than both the p- and q-arms of the respective chromosomes. Additionally, the centromeres for chromosomes 2, 8, and 10 were more distally localized than the q-arm but not the p-arm of the respective chromosomes; albeit the localization of the centromere, p- and q-arm is very similar for chromosome 2.

Table 1. Mean micrometer distance of the geometric center of target loci to the nearest nuclear edge in five subjects for eight different chromosomes.

Chromosome	Distance from the Geometric Center of Target loci to the Nearest Nuclear Edge in μm ± SD			
	p-arm	Centromere	q-arm	Sperm Radius (μm)
2	1.87 ± 0.76	1.87 ± 0.73	1.88 ± 0.75	7.72 ± 0.59
3	1.7 ± 0.69	1.48 ± 0.83	1.68 ± 0.74	7.33 ± 0.83
6	1.93 ± 0.74	1.86 ± 0.83	1.91 ± 0.81	7.7 ± 0.94
8	1.83 ± 0.78	1.94 ± 0.76	1.96 ± 0.82	7.96 ± 0.76
10	1.85 ± 0.83	1.9 ± 0.82	1.96 ± 0.83	7.71 ± 0.73
12	1.89 ± 0.86	1.75 ± 0.86	2.0 ± 0.8	7.57 ± 0.68
16	2.18 ± 0.73	1.97 ± 0.72	1.99 ± 0.73	7.52 ± 0.65
18	1.87 ± 0.84	1.83 ± 0.78	1.86 ± 0.82	7.99 ± 0.86

Table displays the mean micrometer (μm) distance to the nearest nuclear edge of the geometric center of the target loci (centromeres, p- and q-arms) for the investigated chromosomes as determined by the Imaris distance transformation tool. The number of determinants for each chromosome target loci is 155 from five different subjects enrolled in the study. The mean μm distance to the nearest nuclear edge is provided with the standard deviation (SD), as well as the average sperm radius for reference.

3.2. Chromosome p- and q-Arm Configurations in Sperm Nuclei; Is A Hairpin-Loop Formed?

The configurations of each chromosome were assessed by measuring the angle formed by the p- and q-arms through the centromere for each chromosome in the five subjects enrolled in this study. Visualization of the 3D organization of the p- and q-arms of each individual chromosome in sperm nuclei clearly revealed that the p- and q-arms formed distinct territories with little, to no, evidence of intermingling between the two chromosome arms in the vast majority of nuclei (Figures 1 and 2). For the investigated chromosomes we provide evidence to partially support the hairpin-loop model of chromosome organization in human sperm. Our findings clearly demonstrate that the chromosome arms (p and q) had a tendency to lie in parallel to one another forming what can be described as a hairpin-loop configuration (Figures 1 and 2). Analysis of the angles formed by the p- and q-arms through the centromere revealed that when looking at the average of the 155 cells analyzed in the five subjects for each chromosome, the majority of spermatozoa had chromosome hairpin-loop configurations that were less than or equal to 80°. We stratified chromosomes based on the percentage of cells that contained hairpin-loop configurations that were greater than 80°. Between 7–16% of the configurations of chromosomes 2, 8, 10, 12, and 18 were greater than 80°, whereas between 25–38% of the configurations for chromosomes 3, 6, and 16 were greater than 80°. Given that for a significant proportion of chromosomes, the hairpin-loop configurations observed were less than 80°, we further stratified the chromosome configurations to less than or equal to 40° and greater than 40° (Figure 4). The data obtained in this study demonstrates that individual chromosomes preferentially form narrower or wider hairpin-loop configurations that were similar between the five subjects enrolled in this study. The data in Figure 4 displays the chromosomes in order of narrowest (chromosome 10) to widest (chromosome 16) hairpin-loop configuration based on the average data from the five subjects enrolled in this study.

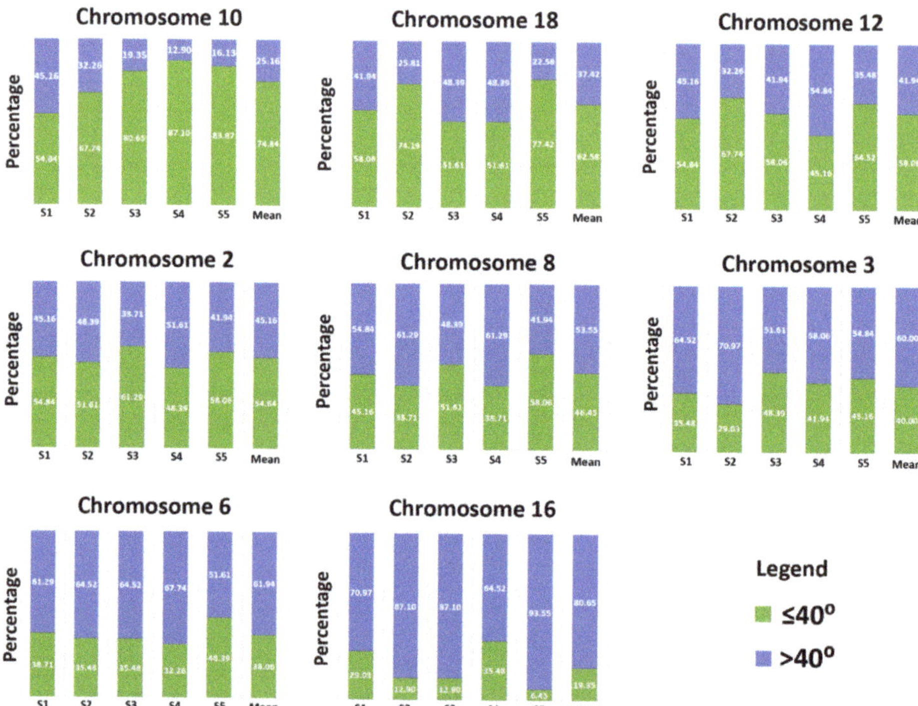

Figure 4. Chromosomes reproducibly have a tendency to form narrower or wider hairpin-loop chromosome configurations in five normozoospermic males for eight investigated chromosomes. The percentage of chromosomes forming narrow (≤ 40°) or wide (> 40°) hairpin-loop configurations for each investigated chromosome is shown for each normozoospermic subject (S1–S5). The mean data for the five subjects is displayed on the far right for each chromosome. Data per subject, per chromosome is based on a minimum of 30 cells; whereas the mean data presented is based on a minimum of 150 cells from the five subjects. Chromosomes are ordered from the top left to the bottom right based on the percentage of narrow (≤ 40°) hairpin-loop configurations for the mean data, with chromosome 6 and 16 preferentially forming the narrowest and widest hairpin-loops, respectively.

Additionally, correlations between chromosome size, gene density and centromere localization and type of hairpin-loop configurations (narrow or wide) were examined. Analysis of the data presented in Figure 4 shows no obvious correlation with between narrow or wide hairpin-loop configuration and chromosome size. For example, chromosome 16 is the second smallest investigated chromosome and preferentially exhibits a wide hairpin-loop configuration, whereas chromosome 18 is the smallest chromosome, and preferentially exhibits a narrow hairpin-loop configuration. Using the Ensembl genome browser correlations between chromosome gene density and hairpin-loop configurations could also be assessed. Chromosome length and number of coding genes for each of the investigated chromosomes was obtained from Ensembl (genome assembly GRCh38.p12). No clear correlation between gene density and tendency to form narrow or wide hairpin loop configurations was observed. For example taking into account the chromosome length and number of protein coding genes, chromosomes 16, 12, 8, 6, 10, 3, 2, and 18 can be organized from highest to lowest gene density respectively. Similar to chromosome size no discernable correlation was observed for gene density and hairpin-loop configuration. For example, looking at the three chromosomes with the highest gene density, they were ranked as third, fifth, and eighth widest hairpin-loop for chromosomes 12, 8, and 16 respectively. Similarly, no correlation was observed between preference to form narrow or wide hairpin-loop configuration with distance of the centromere to the nuclear edge.

4. Discussion

In this study, we investigated the 3D nuclear organization of three major chromosome components (centromeres, p- and q-arms), and how these were configured and formed CTs in sperm nuclei. We targeted eight different metacentric or sub-metacentric chromosomes of varying size and gene density in five normozoospermic males. The results of this study provides further evidence of a non-random radial organization of chromosomes in sperm nuclei, which was reproducible between the enrolled subjects. Of the 24 loci evaluated, all but one (chromosome 12 centromere) were found to be non-randomly radially organized in human sperm. This is not the first study to report the non-random organization of CTs or other target loci; with multiple studies reporting similar observations in humans and evolutionarily divergent species [10,11,14–26,28,32]. Thus, the finding of non-random radial organization for the centromeres, p- and q-arms for the eight investigated chromosomes, is perhaps not surprising. Nevertheless, albeit in a small sample size, one of the major strengths of the current study is that we have examined the radial organization of the various loci between subjects and report a largely reproducible pattern of organization. Interindividual differences are rarely examined in nuclear organization studies, however, a handful of published also report similar findings [10,11,15,19].

CT organization in spermatozoa is hypothesized to possess a unique hairpin-loop structure, which differs from other cell types. The hairpin-loop model proposes that the centromeres form a single or multiple chromocenter in the nuclear interior, with the p- and q-arms of the chromosomes stretching out toward the nuclear periphery where the telomeres form dimers and tetramers [29,31,52]. It is important to note that one of the original experimental papers examining the hairpin-loop conformation in sperm utilized FISH probes against the telomeres, p- and q-arms solely and inferred the position or localization of the centromere based on overlapping signals between the p- and q-arms [29]. The inference of centromere localization is challenging given that this study reported intertwisted spirally-coiled structures and overlapping regions which could lead to erroneous assignment of centromere localization. In the current study, we utilized FISH probes against not only the p- and q-arms but also the centromeres of each investigated chromosome to ensure the CT conformation and localization could be accurately determined for the chromosome arms and centromeres. In support of the previous study, we observed that the p- and q-arms did in fact form a configuration that reflected a hairpin-loop structure, providing further evidence to support the sperm hairpin-loop model of CT organization [25,29–32]. In the current study, the conformation of the p- and q-arms in sperm nuclei were observed to form discrete separate territories with virtually no overlapping or intermingling of the territories for all of the investigated chromosomes, except in rare instances (Figures 1 and 2). Similar findings have also been reported in earlier stages of spermatogenesis [53]; in human lymphocytes and amniotic cells [54] with observed intermingling being limited to a narrow boundary zone. These findings contradict an earlier study that reported the p- and q-arms of three investigated chromosomes (1, 2, and 5) to be tightly bound resulting in intertwisted spirally-coiled structures or closely aligned parallel to each other [29]. Additionally, we observed that certain chromosomes preferentially formed narrow or wide hairpin-loop configurations as determined by the angle created by the arms of each chromosome through the centromere. However, there was no obvious correlation between the angle formed and chromosome size, gene density, or centromere position.

The preference for certain chromosomes to form hairpin-loops that had a tendency to be narrower (e.g., chromosomes 10 and 18) or wider (e.g., chromosome 6 and 16) was largely reproducible between the enrolled subjects, suggesting that this hairpin-loop configuration may be functionally significant.

Despite the aforementioned data, results from this study only partially supports the hairpin-loop model of sperm chromosome organization; given that we were not able to find evidence to support the chromocenter aspect of the model. Rather data from the current study further supports our hypothesis to suggest that the sperm nucleus is reproducibly more segmentally organized than initially hypothesized [6,10]. In line with our previous findings, we observed centromeres to be localized throughout the nucleus (interior, intermediate, and periphery), not just restricted to the nuclear interior. The majority of centromeres (> 65%) were found to be more peripherally localized than one or both of the p- and/or q-arms. These findings are supported by an earlier study that also noted centromeres to be more peripherally organized during earlier stages of spermatogenesis [53]. Of note, the eight chromosome centromeres investigated exhibited the most variation in terms of preferential radial distribution in the three nuclear regions when compared to the chromosome p- and q-arms. In the current study, we established the radial organization based on the localization of the geometric center of each target loci as established by the Imaris software. This approach has been previously utilized in various cell types including sperm [9,10]. This methodology reduces the target loci to a single point (geometric center) regardless of size or shape of the target; and provides the exact radial localization of each probe in the nucleus by measuring the micrometer distance from the geometric center to the nearest nuclear edge in any direction. Given that the centromere has a much smaller and regular shaped territory than the p- and q-arms, one might expect more variation in the radial position of the chromosomes arms when compared to the centromeres. However, we observed the opposite, with the centromeres exhibiting more variation in radial position than the chromosome arms. We hypothesize that this observation supports the hypothesis that CT position in the sperm nucleus may be determined, at least in part, by the centromeres. This may be particularly true if specific centromeres preferentially form the same chromocenters in sperm nuclei. Previously, we identified an average of seven chromocenters in sperm using both 2D and 3D approaches [10]. To determine whether chromocenters were preferentially comprised of the same chromosomes, we co-hybridized multiple FISH probes and examined their co-localization in 3D with each chromocenter. In this study, we utilized three different FISH probes: (1) a pancentromeric probe, staining all centromeres; (2) a nuclear organizing region (NOR) probe, staining regions located in close proximity to the centromeres of the five acrocentric chromosomes (13, 14, 15, 21, and 22); and (3) a centromeric probe that cross hybridized, staining the centromeres of chromosomes 1, 5, and 19 due to sequence homology. Our findings provided indirect evidence to support the hypothesis that specific chromosomes preferentially clustered to form the same chromocenter. The centromere probe for chromosomes 1, 5, and 19, rarely formed the same chromocenter, and the NOR probe typically resulted in three discrete loci in close proximity to different chromocenters. Additionally, the targeted centromeres and NOR probes rarely formed the same chromocenter, suggesting that chromocenters are not composed of random chromosomes, but may consist of specific chromosomes [10]. Thus, there may be a chromosome-specific composition of individual chromocenters that may reflect the hierarchal organization of chromosomes in the sperm nuclei [6,11,19]. This could potentially account for the larger variability observed for the radial position of the centromeres compared to the p- and q-arms in the current study. Another possible explanation for the smaller variability observed in the radial positioning of the p- and q-arms could simply be due to the much larger target, which is certainly more varied in terms of size, shape, and orientation when compared to individual centromeres (Figures 1 and 2). As the geometric center for each target was used to assess the radial position, it is possible that these factors may reduce the variability of the p- and q-arms compared to the much smaller centromere.

One of the concerns for the interpretation of the data in this study and study limitations, is the absolute requirement to artificially decondense sperm nuclei prior to FISH due to the extreme compactness of chromatin in sperm [30]. In order, to help maintain the 3D structure of sperm nuclei a mild formaldehyde fixation step was included to induce DNA-protein crosslinks, which has been shown to aid in maintaining the size and shape of nuclei [54]. The agent DTT used to swell sperm nuclei is analogous to disulphide-reducing glutathione, which is found in the cytoplasm of the oocyte, and thus the swelling of the sperm is believed to mimic the decondensation of the paternal genome that occurs following fertilization [29,55]. However, it is critical to note that differences between studies in fixation and swelling protocols in terms of agents and conditions utilized can make study comparisons challenging and could differentially affect CT configurations. For example, visualization of chromocenter(s) in sperm nuclei might be extremely sensitive to decondensation [6]. A handful of studies have assessed chromocenter formation in human sperm by using pancentromeric FISH probes to visualize all centromeres simultaneously. These studies have observed different numbers of chromocenters in human sperm nuclei, with averages varying from 1-to-11 chromocenters in normozoospermic males [6,14,25,31,52]. The decondensation procedure could also potentially affect the configuration and hairpin-loop structure observed in sperm nuclei, which may in part explain different results between studies. Thus, it is critical that decondensation protocols are clearly specified and that decondensation is carefully monitored to cause minimal disruption to the nuclear organization, whilst ensuring robust FISH efficiency. Consequently, it is important to consider that the CT configurations observed in the decondensed sperm nucleus may not entirely reflect the native CT organization in spermatozoa. The fixation and decondensation methodology utilized in this study resulted in minimal decondensation, high FISH efficiency, and reproducible CT organization in five subjects with comparable findings to a previous study and in a different patient cohort [10]. Thus, we are confident that our approach is consistent and at least reflects the nuclear organization of sperm following the decondensation protocol described.

Differences observed in sperm chromatin organization studies are difficult to resolve and are likely due to inherent differences between studies. For example, there is a large degree of heterogeneity in human sperm samples, and differences may be found between fertile and infertile patients, or may be associated with specific semen parameter disturbances etc. Several studies have reported perturbations, albeit often modest alterations in the positioning of various target loci. Alterations have been reported in patients exhibiting semen parameter disturbances and/or increased sperm aneuploidy [15,19,56,57], increased levels of DNA damage or nuclear perturbations such as sperm nuclear vacuoles [58,59], or carriers of chromosome translocations [45]. There are also inherent methodological differences between studies. These differences include but are not limited to: (1) methodological differences for sperm storage, fixation, and decondensation; (2) differences in the FISH probes utilized; (3) whether 2D or 3D imaging was utilized; and (4) differences in the software and approaches to analyze and determine nuclear organization of probe targets. Thus, direct study comparisons are hampered by these many potential sources of methodological and subject confounders; and are further hindered by the fact that studies often do not adequately describe these important elements to facilitate replication of studies and methodological approaches.

5. Conclusions

Despite the fact, that some published studies are over 20 years old; exploration into the relationship between the organization and function of sperm chromatin organization remains very much in its infancy [6,45]. However, it is clear that spermatozoa possess an evolutionarily conserved unique non-random nuclear organization that differs from somatic cells, suggesting that this organization is functionally significant [16,31]. To date, the function of sperm CT organization remains largely undiscovered [6,44,45]. The current study further supports our hypothesis to suggest that the hairpin-loop model of sperm chromatin organization needs to be refined to reflect a more segmentally organized nucleus. In that, sperm nuclei seemingly do not possess centralized chromocenters, and preferentially form narrower or wider hairpin-loops that can be oriented in any direction in the nucleus, with limited intermingling between chromosome arms. The results suggest that individual chromocenters could be preferentially composed of the same chromosomes, which could determine the segmental radial and longitudinal CT organization, and the hairpin-loop configurations formed. The localization and hairpin-loop configuration of CTs in spermatozoa likely determines the order that CTs or specific genomic regions are exposed to, and remodeled by the oocyte. Thus, nuclear CT position may be critical if there is a functional importance to the order that specific genomic regions are delivered to the oocyte prior to the expression of the paternal genome in the early embryo [6,45]. Therefore, perturbations in sperm CT organization could disrupt the structured events that occur during fertilization, including the formation of the male pronucleus, and early embryonic development [6,14,15,18,29,31,44]. A handful of studies have reported mild to moderate alterations in chromatin organization in the sperm from various patient cohorts including: men with reduced/altered semen parameters, carriers of structural chromosome aberrations, increased sperm aneuploidy and DNA fragmentation etc. [15,19,45,56–59]. However, these studies are restricted to small numbers of patients and cells. Therefore, despite a relationship between chromatin organization, genome regulation and infertility, it is clear that assessment of chromatin organization is far from ready to be implemented into clinical practice [19]. It is possible that CT organization may be an important marker of chromatin integrity that could identify epigenetic perturbations in infertile men [58]. Future studies are required to establish the function of sperm CT organization and its impact on fertilization and early embryonic development. The relationship between sperm nuclear architecture and genome regulation needs to be deciphered, particularly as it pertains to the progression of spermatogenesis, the formation of mature spermatozoa, fertilization, and post-fertilization events. In other words, is organization of the paternal genome important for the development and function of sperm, formation of the male pronucleus, genome activation, and gene expression in the early developing embryo? Ultimately, this may lead to the development of novel clinically relevant tests to assess epigenetic alterations and fertility potential in men, (albeit unlikely to be assessment of CT organization by FISH). Additionally, this could lead to the development of alternative treatment options to those currently available for the treatment of male infertility. In doing so, we may be able to reduce the financial and emotional burden of infertility and assisted reproductive technology, whilst improving the success rates for those couples trying to conceive.

Author Contributions: Conceptualization, D.I. and H.G.T.; Methodology, D.I. and H.G.T.; Validation, D.I.; Formal Analysis, D.I.; Investigation, D.I. and H.G.T.; Writing—Original Draft Preparation, H.G.T.; Writing—Review & Editing, D.I. and H.G.T.; Visualization, D.I. and H.G.T.; Supervision, H.G.T.; Project Administration, D.I.; Funding Acquisition, H.G.T.

Funding: This research received no external funding. The APC was funded by startup funds awarded to HGT from the Herbert Wertheim College of Medicine, Florida International University. The sponsors had no role in the design, execution, interpretation, or writing of the study.

Acknowledgments: This study was made possible through startup funds awarded to HGT from the Herbert Wertheim College of Medicine, Florida International University.

Conflicts of Interest: The authors declare no conflict of interest.

References

1. Fraser, J.; Williamson, I.; Bickmore, W.A.; Dostie, J. An Overview of Genome Organization and How We Got There: From FISH to Hi-C. *Microbiol. Mol. Biol. Rev.* **2015**, *79*, 347–372. [CrossRef] [PubMed]

2. Boveri, T. Die Blastomerenkerne von Ascaris megalocephala und die Theorie der Chromosomenindividualitat. *Arch. Zellforsch.* **1909**, *3*, 181–268.

3. Cremer, T.; Cremer, C. Rise, fall and resurrection of chromosome territories: A historical perspective. Part II. Fall and resurrection of chromosome territories during the 1950s to 1980s. Part III. Chromosome territories and the functional nuclear architecture: Experiments and models from the 1990s to the present. *Eur. J. Histochem.* **2006**, *50*, 223–272. [PubMed]

4. Cremer, T.; Cremer, M. Chromosome territories. *Cold Spring Harb. Perspect. Biol.* **2010**, *2*, a003889. [CrossRef] [PubMed]

5. Meaburn, K.J.; Misteli, T. Cell biology: Chromosome territories. *Nature* **2007**, *445*, 379–781. [CrossRef]

6. Ioannou, D.; Tempest, H.G. Does genome organization matter in spermatozoa? A refined hypothesis to awaken the silent vessel. *Syst. Biol. Reprod. Med.* **2018**, 1–17. [CrossRef] [PubMed]

7. Tanabe, H.; Muller, S.; Neusser, M.; von Hase, J.; Calcagno, E.; Cremer, M.; Solovei, I.; Cremer, C.; Cremer, T. Evolutionary conservation of chromosome territory arrangements in cell nuclei from higher primates. *Proc. Natl. Acad Sci. USA* **2002**, *99*, 4424–4429. [CrossRef]

8. Ioannou, D.; Kandukuri, L.; Quadri, A.; Becerra, V.; Simpson, J.L.; Tempest, H.G. Spatial positioning of all 24 chromosomes in the lymphocytes of six subjects: evidence of reproducible positioning and spatial repositioning following DNA damage with hydrogen peroxide and ultraviolet B. *PLoS ONE* **2015**, *10*, e0118886. [CrossRef]

9. Ioannou, D.; Kandukuri, L.; Simpson, J.L.; Tempest, H.G. Chromosome territory repositioning induced by PHA-activation of lymphocytes: A 2D and 3D appraisal. *Mol. Cytogenet.* **2015**, *8*, 47. [CrossRef]

10. Ioannou, D.; Millan, N.M.; Jordan, E.; Tempest, H.G. A new model of sperm nuclear architecture following assessment of the organization of centromeres and telomeres in three-dimensions. *Sci. Rep.* **2017**, *7*, 41585. [CrossRef]

11. Millan, N.M.; Lau, P.; Hann, M.; Ioannou, D.; Hoffman, D.; Barrionuevo, M.; Maxson, W.; Ory, S.; Tempest, H.G. Hierarchical radial and polar organisation of chromosomes in human sperm. *Chromosome Res.* **2012**, *20*, 875–887. [CrossRef]

12. Foster, H.A.; Bridger, J.M. The genome and the nucleus: a marriage made by evolution. Genome organisation and nuclear architecture. *Chromosoma* **2005**, *114*, 212–229. [CrossRef] [PubMed]

13. Cremer, T.; Kupper, K.; Dietzel, S.; Fakan, S. Higher order chromatin architecture in the cell nucleus: on the way from structure to function. *Biol. Cell* **2004**, *96*, 555–567. [CrossRef] [PubMed]

14. Alladin, N.; Moskovtsev, S.I.; Russell, H.; Kenigsberg, S.; Lulat, A.G.; Librach, C.L. The three-dimensional image analysis of the chromocenter in motile and immotile human sperm. *Syst. Biol. Reprod. Med.* **2013**, *59*, 146–152. [CrossRef]

15. Finch, K.A.; Fonseka, K.G.; Abogrein, A.; Ioannou, D.; Handyside, A.H.; Thornhill, A.R.; Hickson, N.; Griffin, D.K. Nuclear organization in human sperm: preliminary evidence for altered sex chromosome centromere position in infertile males. *Human Reprod. (Oxford, England)* **2008**, *23*, 1263–1270. [CrossRef]

16. Foster, H.A.; Abeydeera, L.R.; Griffin, D.K.; Bridger, J.M. Non-random chromosome positioning in mammalian sperm nuclei, with migration of the sex chromosomes during late spermatogenesis. *J. Cell Sci.* **2005**, *118*, 1811–1820. [CrossRef] [PubMed]

17. Haaf, T.; Ward, D.C. Higher order nuclear structure in mammalian sperm revealed by in situ hybridization and extended chromatin fibers. *Exp. Cell Res.* **1995**, *219*, 604–611. [CrossRef]

18. Hazzouri, M.; Rousseaux, S.; Mongelard, F.; Usson, Y.; Pelletier, R.; Faure, A.K.; Vourc'h, C.; Sele, B. Genome organization in the human sperm nucleus studied by FISH and confocal microscopy. *Mol. Reprod. Dev.* **2000**, *55*, 307–315. [CrossRef]

19. Ioannou, D.; Meershoek, E.J.; Christopikou, D.; Ellis, M.; Thornhill, A.R.; Griffin, D.K. Nuclear organisation of sperm remains remarkably unaffected in the presence of defective spermatogenesis. *Chromosome Res.* **2011**, *19*, 741–753. [CrossRef]

20. Johnson, G.D.; Jodar, M.; Pique-Regi, R.; Krawetz, S.A. Nuclease Footprints in Sperm Project Past and Future Chromatin Regulatory Events. *Sci. Rep.* **2016**, *6*, 25864. [CrossRef]

21. Luetjens, C.M.; Payne, C.; Schatten, G. Non-random chromosome positioning in human sperm and sex chromosome anomalies following intracytoplasmic sperm injection. *Lancet* **1999**, *353*, 1240. [CrossRef]

22. Manvelyan, M.; Hunstig, F.; Bhatt, S.; Mrasek, K.; Pellestor, F.; Weise, A.; Simonyan, I.; Aroutiounian, R.; Liehr, T. Chromosome distribution in human sperm - a 3D multicolor banding-study. *Mol. Cytogenet.* **2008**, *1*, 25. [CrossRef] [PubMed]

23. Zalenskaya, I.A.; Bradbury, E.M.; Zalensky, A.O. Chromatin structure of telomere domain in human sperm. *Biochem. Biophys. Res. Commun.* **2000**, *279*, 213–218. [CrossRef] [PubMed]

24. Zalenskaya, I.A.; Zalensky, A.O. Non-random positioning of chromosomes in human sperm nuclei. *Chromosome Res.* **2004**, *12*, 163–173. [CrossRef] [PubMed]

25. Zalensky, A.O.; Breneman, J.W.; Zalenskaya, I.A.; Brinkley, B.R.; Bradbury, E.M. Organization of centromeres in the decondensed nuclei of mature human sperm. *Chromosoma* **1993**, *102*, 509–518. [CrossRef] [PubMed]

26. Greaves, I.K.; Rens, W.; Ferguson-Smith, M.A.; Griffin, D.; Marshall Graves, J.A. Conservation of chromosome arrangement and position of the X in mammalian sperm suggests functional significance. *Chromosome Res.* **2003**, *11*, 503–512. [CrossRef]

27. Solovei, I.V.; Joffe, B.I.; Hori, T.; Thomson, P.; Mizuno, S.; Macgregor, H.C. Unordered arrangement of chromosomes in the nuclei of chicken spermatozoa. *Chromosoma* **1998**, *107*, 184–188. [CrossRef]

28. Champroux, A.; Damon-Soubeyrand, C.; Goubely, C.; Bravard, S.; Henry-Berger, J.; Guiton, R.; Saez, F.; Drevet, J.; Kocer, A. Nuclear Integrity but Not Topology of Mouse Sperm Chromosome is Affected by Oxidative DNA Damage. *Genes* **2018**, 501. [CrossRef]

29. Mudrak, O.; Tomilin, N.; Zalensky, A. Chromosome architecture in the decondensing human sperm nucleus. *J. Cell Sci.* **2005**, *118*, 4541–4550. [CrossRef]

30. Solov'eva, L.; Svetlova, M.; Bodinski, D.; Zalensky, A.O. Nature of telomere dimers and chromosome looping in human spermatozoa. *Chromosome Res.* **2004**, *12*, 817–823. [CrossRef]

31. Zalensky, A.; Zalenskaya, I. Organization of chromosomes in spermatozoa: An additional layer of epigenetic information? *Biochem. Soc. Trans.* **2007**, *35*, 609–611. [CrossRef] [PubMed]

32. Zalensky, A.O.; Tomilin, N.V.; Zalenskaya, I.A.; Teplitz, R.L.; Bradbury, E.M. Telomere-telomere interactions and candidate telomere binding protein(s) in mammalian sperm cells. *Exp. Cell Res.* **1997**, *232*, 29–41. [CrossRef]

33. Dunleavy, J.E.M.; O'Bryan, M.; Stanton, P.G.; O'Donnell, L. The Cytoskeleton in Spermatogenesis. *Reproduction* **2018**. [CrossRef] [PubMed]

34. Miller, D.; Brinkworth, M.; Iles, D. Paternal DNA packaging in spermatozoa: More than the sum of its parts? DNA, histones, protamines and epigenetics. *Reproduction* **2010**, *139*, 287–301. [CrossRef] [PubMed]

35. Carrell, D.T.; Hammoud, S.S. The human sperm epigenome and its potential role in embryonic development. *Mol. Hum. Reprod.* **2010**, *16*, 37–47. [CrossRef] [PubMed]

36. McLay, D.W.; Clarke, H.J. Remodelling the paternal chromatin at fertilization in mammals. *Reproduction* **2003**, *125*, 625–633. [CrossRef] [PubMed]

37. Arpanahi, A.; Brinkworth, M.; Iles, D.; Krawetz, S.A.; Paradowska, A.; Platts, A.E.; Saida, M.; Steger, K.; Tedder, P.; Miller, D. Endonuclease-sensitive regions of human spermatozoal chromatin are highly enriched in promoter and CTCF binding sequences. *Genome. Res.* **2009**, *19*, 1338–1349. [CrossRef]

38. Hammoud, S.S.; Nix, D.A.; Zhang, H.; Purwar, J.; Carrell, D.T.; Cairns, B.R. Distinctive chromatin in human sperm packages genes for embryo development. *Nature* **2009**, *460*, 473–478. [CrossRef]

39. Martins, R.P.; Krawetz, S.A. Decondensing the protamine domain for transcription. *Proc. Natl. Acad. Sci. USA* **2007**, *104*, 8340–8345. [CrossRef]

40. Wykes, S.M.; Krawetz, S.A. The structural organization of sperm chromatin. *J. Biol. Chem.* **2003**, *278*, 29471–29477. [CrossRef]

41. Jung, Y.H.; Sauria, M.E.G.; Lyu, X.; Cheema, M.S.; Ausio, J.; Taylor, J.; Corces, V.G. Chromatin States in Mouse Sperm Correlate with Embryonic and Adult Regulatory Landscapes. *Cell Rep.* **2017**, *18*, 1366–1382. [CrossRef] [PubMed]

42. Flyamer, I.M.; Gassler, J.; Imakaev, M.; Brandao, H.B.; Ulianov, S.V.; Abdennur, N.; Razin, S.V.; Mirny, L.A.; Tachibana-Konwalski, K. Single-nucleus Hi-C reveals unique chromatin reorganization at oocyte-to-zygote transition. *Nature* **2017**, *544*, 110–114. [CrossRef] [PubMed]

43. Ke, Y.; Xu, Y.; Chen, X.; Feng, S.; Liu, Z.; Sun, Y.; Yao, X.; Li, F.; Zhu, W.; Gao, L.; et al. 3D Chromatin Structures of Mature Gametes and Structural Reprogramming during Mammalian Embryogenesis. *Cell* **2017**, *170*, 367.e20–381.e20. [CrossRef] [PubMed]

44. Sarrate, Z.; Sole, M.; Vidal, F.; Anton, E.; Blanco, J. Chromosome positioning and male infertility: it comes with the territory. *J. Assisted Reprod. Genet.* **2018**, *35*, 1929–1938. [CrossRef] [PubMed]

45. Wiland, E.; Zegalo, M.; Kurpisz, M. Interindividual differences and alterations in the topology of chromosomes in human sperm nuclei of fertile donors and carriers of reciprocal translocations. *Chromosome Res.* **2008**, *16*, 291–305. [CrossRef] [PubMed]

46. Wu, T.F.; Chu, D.S. Epigenetic processes implemented during spermatogenesis distinguish the paternal pronucleus in the embryo. *Reprod. Biomed. Online* **2008**, *16*, 13–22. [CrossRef]

47. World Health Organization. *WHO Laboratory Manual for the Examination and Processing of Human Semen*, 5th ed.; World Health Organization: Geneva, Switzerland, 2010; p. 287.

48. Acar, M.; Kocherlakota, K.S.; Murphy, M.M.; Peyer, J.G.; Oguro, H.; Inra, C.N.; Jaiyeola, C.; Zhao, Z.; Luby-Phelps, K.; Morrison, S.J. Deep imaging of bone marrow shows non-dividing stem cells are mainly perisinusoidal. *Nature* **2015**, *526*, 126–130. [CrossRef]

49. Takaku, T.; Malide, D.; Chen, J.; Calado, R.T.; Kajigaya, S.; Young, N.S. Hematopoiesis in 3 dimensions: human and murine bone marrow architecture visualized by confocal microscopy. *Blood* **2010**, *116*, e41–e55. [CrossRef]

50. Whalen, K.; Reitzel, A.M.; Hamdoun, A. Actin polymerization controls the activation of multidrug efflux at fertilization by translocation and fine-scale positioning of ABCB1 on microvilli. *Mol. Biol. Cell* **2012**, *23*, 3663–3672. [CrossRef]

51. Cukierski, W.J.; Nandy, K.; Gudla, P.; Meaburn, K.J.; Misteli, T.; Foran, D.J.; Lockett, S.J. Ranked retrieval of segmented nuclei for objective assessment of cancer gene repositioning. *BMC Bioinformatics* **2012**, *13*, 232. [CrossRef]

52. Zalensky, A.O.; Allen, M.J.; Kobayashi, A.; Zalenskaya, I.A.; Balhorn, R.; Bradbury, E.M. Well-defined genome architecture in the human sperm nucleus. *Chromosoma* **1995**, *103*, 577–590. [CrossRef] [PubMed]

53. Scherthan, H.; Eils, R.; Trelles-Sticken, E.; Dietzel, S.; Cremer, T.; Walt, H.; Jauch, A. Aspects of three-dimensional chromosome reorganization during the onset of human male meiotic prophase. *J. Cell Sci.* **1998**, *111 (Pt 16)*, 2337–2351.

54. Dietzel, S.; Jauch, A.; Kienle, D.; Qu, G.; Holtgreve-Grez, H.; Eils, R.; Munkel, C.; Bittner, M.; Meltzer, P.S.; Trent, J.M.; et al. Separate and variably shaped chromosome arm domains are disclosed by chromosome arm painting in human cell nuclei. *Chromosome Res.* **1998**, *6*, 25–33. [CrossRef] [PubMed]

55. Sutovsky, P.; Schatten, G. Depletion of glutathione during bovine oocyte maturation reversibly blocks the decondensation of the male pronucleus and pronuclear apposition during fertilization. *Biol. Reprod.* **1997**, *56*, 1503–1512. [CrossRef] [PubMed]

56. Abdelhedi, F.; Chalas, C.; Petit, J.M.; Abid, N.; Mokadem, E.; Hizem, S.; Kamoun, H.; Keskes, L.; Dupont, J.M. Altered three-dimensional organization of sperm genome in DPY19L2-deficient globozoospermic patients. *J. Assisted Reprod. Genet.* **2019**, *36*, 69–77. [CrossRef] [PubMed]

57. Olszewska, M.; Wiland, E.; Kurpisz, M. Positioning of chromosome 15, 18, X and Y centromeres in sperm cells of fertile individuals and infertile patients with increased level of aneuploidy. *Chromosome Res.* **2008**, *16*, 875–890. [CrossRef] [PubMed]

58. Perdrix, A.; Travers, A.; Clatot, F.; Sibert, L.; Mitchell, V.; Jumeau, F.; Mace, B.; Rives, N. Modification of chromosomal architecture in human spermatozoa with large vacuoles. *Andrology* **2013**, *1*, 57–66. [CrossRef] [PubMed]

59. Wiland, E.; Fraczek, M.; Olszewska, M.; Kurpisz, M. Topology of chromosome centromeres in human sperm nuclei with high levels of DNA damage. *Sci. Rep.* **2016**, *6*, 31614. [CrossRef]

 *genes*

Article

Telomere Dynamics Throughout Spermatogenesis

Heather E. Fice [1] and Bernard Robaire [1,2,*

1   Department of Pharmacology and Therapeutics, McGill University, Montreal, QC H3G 1Y6, Canada
2   Departments of Obstetrics and Gynecology, McGill University, Montreal, QC H4A 3J1, Canada
*   Correspondence: bernard.robaire@mcgill.ca; Tel.: +1-514-398-3630

Received: 14 May 2019; Accepted: 10 July 2019; Published: 12 July 2019

Abstract: Telomeres are repeat regions of DNA that cap either end of each chromosome, thereby providing stability and protection from the degradation of gene-rich regions. Each cell replication causes the loss of telomeric repeats due to incomplete DNA replication, though it is well-established that progressive telomere shortening is evaded in male germ cells by the maintenance of active telomerase. However, germ cell telomeres are still susceptible to disruption or insult by oxidative stress, toxicant exposure, and aging. Our aim was to examine the relative telomere length (rTL) in an outbred Sprague Dawley (SD) and an inbred Brown Norway (BN) rat model for paternal aging. No significant differences were found when comparing pachytene spermatocytes (PS), round spermatids (RS), and sperm obtained from the caput and cauda of the epididymis of young and aged SD rats; this is likely due to the high variance observed among individuals. A significant age-dependent decrease in rTL was observed from 115.6 (±6.5) to 93.3 (±6.3) in caput sperm and from 142.4 (±14.6) to 105.3 (±2.5) in cauda sperm from BN rats. Additionally, an increase in rTL during epididymal maturation was observed in both strains, most strikingly from 115.6 (±6.5) to 142 (±14.6) in young BN rats. These results confirm the decrease in rTL in rodents, but only when an inbred strain is used, and represent the first demonstration that rTL changes as sperm transit through the epididymis.

Keywords: male germ cells; spermatogenesis; sperm; telomeres; chromatin; reproductive aging

---

1. Introduction

The male germline is biologically unique in many ways, ranging from cellular structures to chromatin packaging and enzymatic activity. Telomeres are no exception to this statement, with telomere dynamics in male germ cells being distinctly different from those of somatic cells. Telomeres are 5′-TTAGGG-3′ repeat sequences that cap the ends of chromosomes to give the genome protection and stability from progressive shortening after DNA replication caused by the incomplete replication of the 5′ end by DNA polymerase [1]. The progressive shortening of telomeres due to the end-replication problem can be mitigated by the enzyme telomerase; it maintains telomeres by the addition of new repeats. Containing both a protein (TERT) and an RNA template (TERC), the enzyme functions as a reverse transcriptase synthesizing a single strand of telomeric DNA complementary to TERC onto the 3′ overhang. This newly synthesized telomeric DNA strand is then used for lagging strand synthesis by DNA replication machinery [1]. Alternative lengthening of telomeres (ALT) by homologous recombination is another mechanism by which telomeres and subtelomeric regions are able to retain their length in the absence of active telomerase [2].

The existing literature presents data on the length and spatial arrangement of telomeres, and the activity of telomerase within germ cell nuclei as spermatogenesis progresses [3]. Beginning in spermatogonia, telomerase is most active and telomere length is hypothesized to be shorter relative to fully mature spermatozoa [4]. The telomeres at this stage are randomly positioned; however, during mitosis, they align to either pole of the cell in preparation for cytokinesis. In spermatocytes, telomerase levels are high and telomeres follow a similar alignment once meiotic events are initiated. Round spermatids have similarly high levels of telomerase at the onset of spermiogenesis; however, these levels decrease as the cells become transcriptionally inactive during chromatin compaction [4]. At this stage, fluorescence in situ hybridization (FISH) experiments have also shown that telomeres spread randomly throughout the cell nucleus. Interestingly though, FISH experiments often display a reduced number of telomeres due to their apparent dimerization. This has been shown as the number of telomeres present at the final stages of spermiogenesis is half of the expected number, suggesting that they are co-localizing [5–7]. Some hypotheses have been put forth about the nature of this interaction, and through FISH experimental staining for p and q arms of chromosomes 3 and 6, it appears that the telomeres of each chromosome bind to each other in a loop-like fashion [7]. Throughout spermiogenesis and epididymal transit, germ cells undergo dramatic chromatin repackaging. There is a gradual replacement of most histone-bound nucleosomes first with transition proteins and then with protamines, to form a tight toroidal conformation [8]. This repackaging event does not completely void the cell of histones, and approximately 10–15% of histones are retained in human sperm [9–11], while in rodents, only 1–2% of histones are retained [11,12]. Fully mature sperm maintain dimeric telomeres, as shown in round spermatids, though they contrast with earlier germ cells as little to no telomerase activity has been observed [13]. Telomere length in spermatozoa is longer than in somatic cells and has been measured at approximately 6–20 kb in humans [14–18]. Spermatozoa also appear to have a specific organization of telomeres, with the telomeric regions of chromatin found toward the nuclear periphery or bound to the nuclear membrane. This observation has been shown for many species, including humans, rodents, primates, and bovine [14,19]. It has been postulated that the combination of histone-bound telomeres and their arrangement at the nuclear periphery serves a functional role after fertilization as these sites are more readily accessible by the oocyte for pronuclear formation [6,20].

A central current issue in male germ cell telomere biology is whether telomere length can be used as a biomarker for sperm quality and fertility. The parameters set by the World Health Organization (WHO) used to assess male fertility do not capture information about sperm chromatin quality [21]. Although measuring sperm DNA integrity is considered an important endpoint [22], many of the methods have been classically challenging in a clinical setting as they require a high level of technical expertise. As a result, there is a demand for a quick reproducible test that would examine a new sperm parameter. Telomere length is a desirable measure, as preliminary studies are beginning to suggest links between fertility outcomes and sperm telomere length. However, there is some controversy in this field regarding which methods measure telomere length in a reliable and accurate way. The methods employed include Southern blotting, fluorescence in situ hybridization, and the quantitative polymerase chain reaction (qPCR). Both Verhulst [23] and Eisenberg [24] have discussed the issues as they relate to each method's reliability, pointing out the inherent cost–benefit analysis that must be done when deciding on a method. When assessing telomere length as a biomarker for fertility in humans, it would be most appropriate to use qPCR as it is relatively simple, inexpensive, and allows for a high-throughput analysis of many samples.

As previously mentioned, preliminary data on the links between sperm telomere length and well-established fertility parameters are beginning to emerge [25]. Several studies have found an association between a shorter telomere length and infertility or oligozoospermia [26–30], but not with classical WHO semen parameters. Interestingly, Garolla et al. found a positive association between sperm telomere length and protamination status [31]. This finding suggests that an error in chromatin packaging results in telomere dysregulation in mature sperm. Additionally, more loosely packaged chromatin could result in an increase in exposure to reactive oxygen species.

There are many factors known to increase male factor infertility, including smoking, alcohol, toxicant exposure, and being overweight [32]. These lifestyle factors, in addition to the aging process, greatly increase the presence of reactive oxygen species; several studies have found an association between these lifestyle factors and disrupted sperm telomere integrity [33,34]. Telomeres are particularly susceptible to oxidative damage as they are highly rich in guanine, allowing for the oxidization to

8-oxo-2′-deoxyguanosine (8-oxo-dG) [35]. In vitro results suggest that oxidative insult results not only in disrupted telomere integrity, but also in telomere shortening [36]. Additionally, the retention of histones in telomeric regions makes these regions more sensitive to oxidative insult [20]. The DNA damage that may be incurred from these oxidative insults can further lead to telomeric instability and telomere–telomere interactions may be lost [37].

Telomere length decreases in somatic cells with advanced age, but there are varying species-dependent effects on sperm telomere length. In studies examining telomere length in mice, the trend with advanced paternal age is a decrease in telomere length, similar to that seen in somatic cells [38]. However, when similar studies were done using human sperm, the telomere length appeared to increase with age [16,39]. There are two main hypotheses addressing the potential cause of telomere lengthening in species with longer life spans. The first is that because telomerase is active in spermatogonia and throughout spermatogenesis, it has ample time to act and build on telomeres as the pool of stem cells is aging. The second is that there is a selection of germ cells for those with the longest telomeres over the course of a man's lifespan, resulting in only those with long telomeres remaining at an advanced age [40].

Telomere homeostasis may exist, where there is a balance for the optimal telomere length. When the telomeres are dysregulated, meiosis can be more error-prone, with chromosome segregation being incomplete and higher rates of aneuploidy [41]. Supporting this hypothesis, Cariati et al. have shown data that there are pregnancy failures when male partners have short telomeres [28]. It is also interesting to note that these studies have explored the association between sperm telomere length and offspring leukocyte telomere length. Few studies have directly studied both sperm telomere length and offspring telomere length; however, in rodents, birds, primates, and humans, there is a clear paternal age effect on telomere length, where older fathers produce offspring with longer telomeres [40,42–45]. These results are in favour of the hypothesis that telomeres are an epigenetic feature.

Although we are gaining insight into several aspects of the length of telomeres in the context of male reproduction, no study to date has related the effects of the phase of spermatogenesis and epididymal sperm maturation to telomere length with advancing paternal age, or established whether observed differences can be accounted for by the use of inbred and outbred rodent strains.

2. Materials and Methods

2.1. Animals

All studies were conducted on Brown Norway (BN) and Sprague Dawley (SD) transgenic rat strains bred in-house, with initial breeding pairs kindly provided by Dr. Hamra at UT Southwestern. The rats were transgenic for td-Tomato red (BN) and e-GFP (SD) expression in the germline. All animals had access to food and water ad libitum, and were kept in a 12-hour light, 12-hour dark, temperature- and humidity-controlled environment. BN and SD rats (n = 3–5) were sacrificed at young and aged time points. The average ages for the inbred BN rats were 5.6 months ± 0.2 and 19.2 ± 0.06 months for young and aged populations, respectively. For outbred SD rats, the average ages were 5.6 ± 0.18 and 18.7 ± 0.32 months for young and aged populations, respectively. Eighteen to twenty months of age in a rat is the age prior to germ cell loss and testicular atrophy [46]. Animal care and handling were done in accordance with the guidelines put forth by the Canadian Council on Animal Care (McGill Animal Resources Centre protocol 4687).

2.2. Germ Cell Separation

Young and aged rats were euthanized by CO_2 asphyxiation. Testes were removed and weighed to assess the regression status associated with advanced aging, and rat testes less than 1.5 grams were considered regressed and not used in this study. When a testis was excluded, the attached epididymis was not used for sperm collection. Of 11 aged animals, four possessed only one testis that was not regressed. No animals had both testes regressed. Germ cells were obtained using the STA-PUT method for cell velocity sedimentation [47]. Briefly, testes were decapsulated prior to enzymatic digestion with 0.5 mg mL^{-1} collagenase (C9722-50MG; Sigma Aldrich, Oakville, Canada), followed by subsequent digestion with 0.5 mg mL^{-1} trypsin (Type I, T8003; Sigma-Aldrich, Oakville, Canada) and DNase I (Type I, DN-25; Sigma-Aldrich, Oakville, Canada). The dissociated germ cell suspension was then filtered through a 70 µM nylon mesh before being washed three times with 0.5% bovine serum albumin (A4612; Sigma Aldrich, Oakville, Canada) in RPMI 1640 (Life Technologies, Grand Island, USA) and pelleted at 233 g for 5 min. Cells were filtered once more with a 55 µM mesh to prevent clumping and 5.5×10^8 mixed germ cells in 25 mL of 0.5% BSA in RPMI were loaded into the STA-PUT (Proscience, Toronto, Canada) and separated on a gradient of 2–4% BSA/RPMI. The gradient was established over 50 min, and the cells were separated through unit gravity sedimentation for 1 h 45 min. A fraction collector was then used to collect the germ cells in individual populations of pachytene spermatocytes (PS), and round (RS) and elongating spermatids. Fractions that met at least 80% purity by phase-contrast microscopy identification were spun down, flash frozen, and kept at −80 °C for future experiments. Spermatozoa from the caput and cauda epididymidis were isolated in PBS after 2 h of agitation. They were filtered through a 100 µM nylon mesh before being centrifuged and washed six times with 0.45% saline solution.

2.3. Telomere Measurement

DNA was extracted from 1.5×10^6 PS, RS, and spermatozoa from the caput and cauda epididymidis using the QiaAMP DNA mini kit (51304; Thermo Fisher Scientific, Mississauga, Canada), with the substitution of a separate sperm lysis buffer including 40 mM dithiothreitol (DTT). Extracted sperm DNA may have different recoverability at the telomeres, as in these regions, it is packaged primarily with histones, while the remainder of the DNA is bound to protamine. Given our protocol for sperm DNA extraction, which disrupts the bound protamines, we did not anticipate that this would be an issue. DNA was diluted to a working concentration of 5 ng µL^{-1} for telomere measurement by qPCR [48] for telomeric repeats and 36B4 single copy gene amplification measured by ΔCt. The mastermix for a final reaction volume of 20 µL per well was prepared using 10 µL per reaction SYBR Green MM solution (4367659; Thermo Fisher Scientific, Mississauga, Canada). For each 36B4 reaction, 1 µL of 2 µM forward and reverse primers for 36B4 and 5 µL PCR grade water were used. For each telomeric DNA reaction, 0.5 µL 2 µM forward and reverse primers for telomeric DNA with 4.5 µL of PCR grade water were used (Table S1). For all reactions, 20 ng/well DNA was used. The standard curve for telomeric repeats follows a 1:5 dilution, beginning with 4000 picograms (pg) of telomere oligomer (Table S2), corresponding to 7.6×10^9 kb. The 36B4 standards begin with a concentration of 2 pg (Table S2), following a 1:10 dilution, corresponding to 3.6×10^9 genome copies. Standards were brought to a total of 20 ng of DNA by spiking with pBR322 DNA. All samples presented herein fall along the presented standard curves. A four-step PCR amplification protocol was used. First, denaturation occurred at 95 °C for ten minutes (one cycle), followed by 40 cycles of denaturation at 95 °C for 15 s and annealing at 60 °C for 1 min. The melt curve conditions were 95 °C for 15 s, and annealing at 60 °C for 1 min, with a temperature increase of 0.5 °C per cycle to 95 °C for 15 s. The final step was an infinite hold at 4 °C. By taking the telomeric repeats, relative to the genome copies, the telomere length per genome was represented. DNA taken from H1301 cells (#01051619-DNA-5UG; Sigma-Aldrich, Oakville, Canada) with a known telomere length of 70 kb was then used for the normalization of all samples.

The calculations are as follows:

- Calculating telomeric repeats on a log scale:

$$\log(Tel) = \frac{\Delta Ct - B}{m} \tag{1}$$

where the telomere standard curve produces the following slope: $\Delta Ct = m \, (logTel) + B$.

- Calculating genome copies (GC) represented by single copy gene 36B4:

$$\log(GC) = \frac{\Delta Ct - B}{m} \tag{2}$$

where the 36B4 standard curve produces the following slope: $\Delta Ct = m\ (logGC) + B$.

- Calculating telomeric repeats per genome (telomere/single copy gene):

$$\log(telomeric\ repeats\ per\ genome) = \frac{\log(Tel)}{\log(GC)} \tag{3}$$

$$telomeric\ repeats\ per\ genome = \log(telomeric\ repeats\ per\ genome)^{10} \tag{4}$$

- Calculating telomere length relative to H1301 cell DNA, with a predicted telomere length of 70 kb:

$$70\ kb = \frac{telomeric\ repeats\ pet\ genome\ H1301}{x} \tag{5}$$

$$relative\ telomere\ length = \frac{telomeric\ repeats\ per\ genome}{x} \tag{6}$$

All experiments were done in triplicate, with intra-class correlation coefficients of 0.82 and 0.85 for young and aged BN sperm telomere lengths, respectively. To control for inter-plate variation, H1301 and standard curves were run on each plate. An inherent limitation of this protocol is the normalization of samples to H1301, as different methods of DNA extraction and handling can alter the apparent measure of telomere length. Though this was controlled for with samples processed in house, H1301 DNA was extracted and purified by Sigma.

2.4. Statistical Analysis

To calculate the telomere length, telomere kb and 36B4 genome copies were extrapolated from the standard curves and ΔCt values (Equations (1) and (2)). The telomere kb was divided by the genome copies represented by 36B4 (Equation (3)). These values were then normalized to the positive control H1301 DNA (Equation (4)), with a known telomere length of 70 kb, to give a measurement of relative telomere length (rTL). The median and interquartile range were calculated in Excel. Further statistics and data analysis were conducted using Graph-Pad Prism 6. Where appropriate, t-tests were used for statistical comparisons between groups; however, where variances were significantly different, a Mann Whitney U test was used as a replacement. Statistical significance of $p \leq 0.05$ has been indicated with an asterisk (*).

3. Results and Discussion

3.1. Telomere Dynamics Show Rat Strain Specificity Between Brown Norway and Sprague Dawley Rats

Telomere length for the outbred SD rats is in the range of 200–350 across spermatogenesis, while that for the inbred BN rats is shorter and has a decreased range of 115–160. Comparative studies of germ cell telomere length across varying species and strains have not been conducted. Although one would anticipate less variance in the lengths of telomeres from an inbred than outbred strain due to decreased genetic heterogeneity, it has also been proposed that inbred strains may have shorter somatic telomeres due to the increased oxidative stress and reduced evolutionary fitness [49]. The fact that this trend is maintained in the germline reveals potential long-term effects in an inbred rat strain as sperm telomere length is correlated with offspring telomere length [40,42–45]. Both strains show no difference in PS or RS rTL, a trending decrease in the caput sperm, and the subsequent recovery of telomere length in sperm from the cauda epididymidis. The most striking difference between strains is that both the interquartile range (IQR) and standard errors calculated for BN sperm telomere lengths are much smaller than those for the SD sperm (Table 1). The interquartile range represents the spread of data, by showing where 50% of the data points lie in a given sample set. The smaller IQR values for BN rats are likely due to the inbred nature of BN rats and the level of their genetic similarity. The homogeneity in rTL further validates them as a model for epigenetic studies in rodents. With both the inherent variability seen in the SD telomere length measurements and the exclusive use of BN rats for epigenetic studies, data for BN rats will be presented throughout the remainder of the text. The SD data is presented in Figure A1.

Table 1. Species variation for telomere length measurement in sperm.

Rat Strain	Cell Type	N	Median rTL	IQR	SEM
SD - Young	PS	5	205.89	126.14	37.21
	RS	4	180.53	141.46	124.19
	CP	6	230.47	79.17	25.54
	CD	10	396.81	253.57	54.94
SD - Aged	PS	6	204.92	132.51	38.99
	RS	5	301.42	122.45	52.70
	CP	3	116.93	205.51	133.00
	CD	12	302.82	132.82	28.49
BN - Young	PS	5	143.70	13.35	11.63
	RS	5	165.79	77.18	20.09
	CP	5	116.61	17.56	6.49
	CD	5	129.67	17.42	14.61
BN - Aged	CP	4	97.47	13.40	6.35
	CD	4	106.54	4.64	2.52

Species differences in relative telomere length (rTL) variability shown between Sprague Dawley (SD) and Brown Norway (BN) rats for pachytene spermatocytes (PS), round spermatids (RS), and sperm taken from the caput (CP) and cauda (CD) epididymis for both young and aged samples. N: Sample Size. IQR: Interquartile Range. SEM: Standard Error of the Mean.

3.2. Telomere Lengths During Spermatogenesis in Brown Norway Rats

Examining germ cell telomere dynamics has been done extensively in the context of telomerase activity, with a well-defined pattern of high telomerase activity in early germ cells that tapers off as spermatogenesis progresses. However, the existing literature that examines telomere length is less complete, mainly examining fully mature sperm and operating under the assumption that germ cell telomere length is strongly correlated with telomerase activity. When measuring rTL in BN rats, we find that there is no significant difference in the telomere length from PS to RS, with lengths measured at 155.4 (±11.6) and 159.2 (±20.1), respectively (Figure 1). This observation suggests that the length of telomeres remains relatively constant throughout the meiotic stages of spermatogenesis, independent of the apparent increase in telomerase activity [4]. An important component of understanding telomere dynamics throughout spermatogenesis that is missing is the measurement of telomere length in the spermatogonial stem cells; however, a methodology for the isolation of rat spermatogonial stem cells has yet to be developed.

Interestingly, when entering the epididymis, the length of telomeres shows a decrease of approximately 25% from what is observed for earlier stages of spermatogenesis. The length of telomeres from the spermatozoa obtained from the caput epididymidis of any species has not been measured previously, so it is difficult to determine if this novel observation can be generalized beyond the rat. This finding suggests altered telomere organization during chromatin condensation and crosslinking through epididymal maturation. However, what is apparent is that by the time sperm reach the cauda epididymidis, the sperm telomere length reaches a length of 142 (±14.6), comparable to the germ cell telomere length prior to entering the epididymis (Figure 1). The epididymis is a tissue that has received relatively little attention; understanding how the environment of the caput, corpus, and cauda epididymidis alters sperm chromatin is a major challenge that needs to be addressed by the scientific community. It is possible that telomere organization is impacted by micro and non-coding RNAs that are passed to the sperm through epididymosomes [50,51]. As more interactions are being elucidated for non-coding RNAs and telomeric regions, the functional role of these interactions will become clearer [52]. Telomeric repeat containing RNA (TERRA) is a non-coding RNA transcribed from telomeric regions that is able to bind telomeric DNA. The proposed function of TERRA binding is to control telomere structure and elongation; this has been shown in various species [53–56]. TERRA has also been shown to modify polycomb repressive complex binding, and modify histone marks across the genome and in telomeres [57]. Though there is limited literature on TERRA in male germ cells, Reig-Viader et al. have shown that it is present in spermatocytes and spermatids [58]. They have also shown that telomeres and TERRA levels were disrupted in germ cells from men with idiopathic infertility [59]. Taken together, these observations indicate the need for further studies to resolve the effects of non-coding RNAs during epididymal maturation.

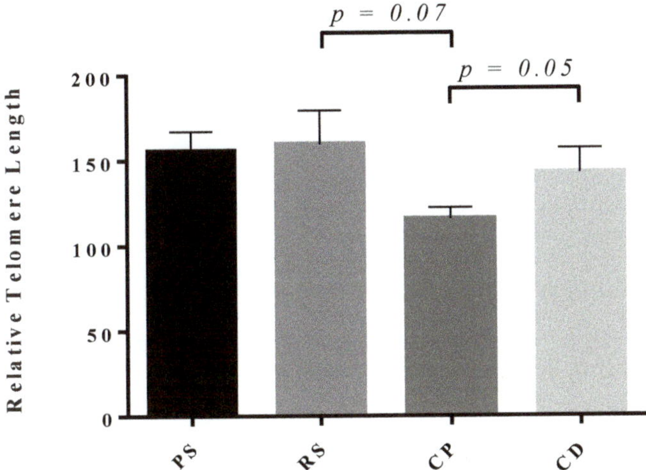

Figure 1. Telomere length in the young Brown Norway male germline. Relative telomere length (rTL) shown on the y-axis measured by quantitative polymerase chain reaction (qPCR) relative to H1301 cell DNA of a known telomere length, for pachytene spermatocytes (PS), round spermatids (RS), caput sperm (CP), and cauda sperm (CD). Each bar represents the mean ± SEM, n = 5. Sprague Dawley data shown in Figure A1.

3.3. Age-Dependent Decrease in Sperm Telomere Length

There is a significant age-dependent decrease in rTL from 115.6 (±6.5) to 93.3 (±6.3) in caput sperm (p = 0.04), which remained consistent for cauda sperm, with a decrease observed from 142.4 (±14.6) to 105.3 (±2.5) in cauda sperm (p = 0.02; Figure 2). This decrease is consistent with mouse models of paternal aging presented in the literature [38]. Interestingly, the trend for increased telomere length during epididymal transit is seemingly reduced with aging. A modest increase in rTL is observed from 93.3 (±6.3) to 105.3 (±2.5) in the caput sperm. If, during epididymal transit, non-coding RNAs contribute to affecting telomere length, it is possible that the epididymosome payload changes with advancing age, though no study to date exists on epididymosomes and aging.

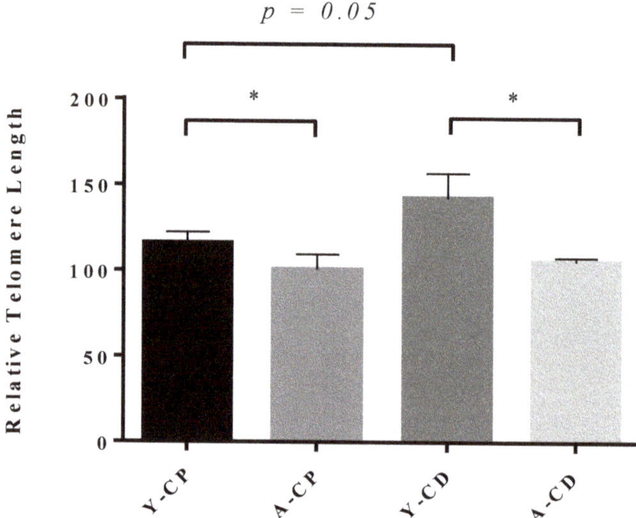

Figure 2. Telomere length for Brown Norway sperm during aging. Relative telomere length (rTL) shown on the y-axis measured by quantitative polymerase chain reaction (qPCR) relative to H1301 cell DNA of a known telomere length, for young caput sperm (Y-CP; n = 5), aged caput sperm (A-CP; n = 4), young cauda sperm (Y-CD; n = 5), and aged cauda sperm (A-CD; n = 4). Each bar represents the mean ± SEM. $p \leq 0.05$ is indicated by an asterisk. Sprague Dawley data shown in Figure A1.

There are currently no hypotheses to address the decrease in telomere length observed in rodent models of paternal aging. However, it seems probable that hypotheses proposed to explain germ cell telomere lengthening in humans may not apply to the much shorter lifespan of a rodent.

4. Conclusions

Understanding telomere length in the varying contexts that influence male reproductive function and spermatogenesis is critical to understanding their epigenetic implications. As telomeres are associated with the nuclear envelope, it remains plausible that they are sites initially recognized by the egg after fertilization to aid in chromatin anchoring; telomere length may also influence offspring health in this way [20]. Altered telomere length, either increased or decreased, may lead to a disruption in chromatin reorganization events following fertilization [28]. Studies by our group have shown several effects of aging on male reproductive outcomes, including increased time to pregnancy, higher resorption rates, and an increased instance of infertility [46]. It is difficult to conclude if the negative outcomes are associated with one specific pathology of aging, such as telomere length, as these cells are also exposed to increased oxidative stress and decreased DNA damage repair, and thus show increased DNA damage. The presence of increased DNA damage with aging has not been examined within telomeric regions; however, it may provide additional insight into sperm telomere dynamics during aging. Here, we have shown that sperm telomere length decreases with age in inbred Brown Norway rats. This poses an interesting question, and by examining telomere dynamics in embryos fertilized with young and aged sperm, we may begin to understand this relationship more clearly. Additionally, using RNA sequencing and chromatin conformation capture methods will elucidate how telomere dynamics are altered across spermatogenesis with aging.

Supplementary Materials: The following are available online at http://www.mdpi.com/2073-4425/10/7/525/s1: Table S1: Telomere Length Quantitative Polymerase Chain Reaction (qPCR) Primers, Table S2: Telomere Length Quantitative Polymerase Chain Reaction (qPCR) Oligomer Standard Sequences.

Author Contributions: Conceptualization, B.R. and H.E.F.; methodology, H.E.F.; investigation, H.E.F.; resources, B.R.; data curation, H.E.F.; writing—original draft preparation, H.E.F.; writing—review and editing, B.R.; supervision, B.R.; project administration, B.R.; funding acquisition, B.R.

Funding: This research has been funded by the CIHR Institute for Gender and Health Team Grant TE1-138298, and HEF has a trainee award provided by the Centre for Research in Reproduction and Development.

Acknowledgments: The authors would like to thank Trang Luu, Aimee Katen, and Anne Marie Downey for their assistance with the development of methods. We would also like Kent Hamra for providing breeder rat pairs for his transgenic lines.

Conflicts of Interest: The authors declare no conflict of interest.

Appendix A

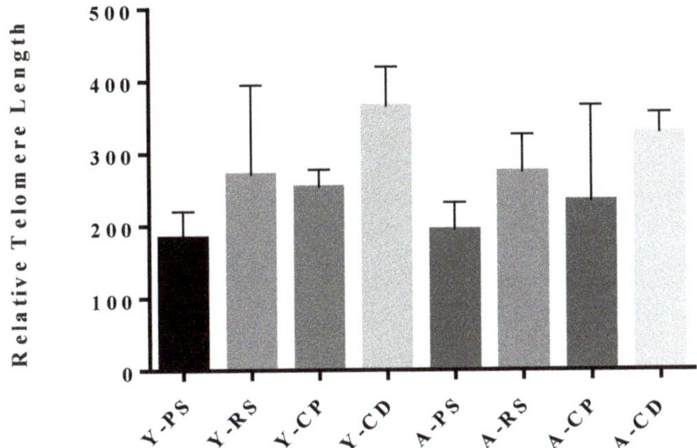

Figure A1. Relative telomere length (rTL) for the Sprague Dawley germline during aging. rTL for PS: Pachytene Spermatocyte, RS: Round Spermatid, CP: Caput Sperm, and CD: Cauda Sperm. Young samples presented as Y- (cell type) and aging samples presented as A- (cell type).

References

1. Chan, S.R.W.L.; Blackburn, E.H. Telomeres and Telomerase. *Philos. Trans. R. Soc. Lond. B Biol. Sci.* **2004**, *359*, 109–121. [CrossRef] [PubMed]

2. Cesare, A.J.; Reddel, R.R. Alternative Lengthening of Telomeres: Models, Mechanisms and Implications. *Nat. Rev. Genet.* **2010**, *11*, 319–330. [CrossRef] [PubMed]

3. Rocca, M.S.; Foresta, C.; Ferlin, A. Telomere Length: Lights and Shadows on Their Role in Human Reproduction. *Biol. Reprod.* **2018**, *100*, 305–317. [CrossRef] [PubMed]

4. Achi, M.V.; Ravindranath, N.; Dym, M. Telomere Length in Male Germ Cells Is Inversely Correlated with Telomerase Activity. *Biol. Reprod.* **2005**, *63*, 591–598. [CrossRef] [PubMed]

5. Zalensky, A.O.; Tomilin, N.V.; Zalenskaya, I.A.; Teplitz, R.L.; Bradbury, E.M. Telomere-Telomere Interactions and Candidate Telomere Binding Protein(s) in Mammalian Sperm Cells. *Exp. Cell Res.* **1997**, *232*, 29–41. [CrossRef] [PubMed]

6. Zalenskaya, I.A.; Zalensky, A.O. Non-Random Positioning of Chromosomes in Human Sperm Nuclei. *Chromosome Res.* **2004**, *12*, 163–173. [CrossRef]

7. Solov'eva, L.; Svetlova, M.; Bodinski, D.; Zalensky, A.O. Nature of Telomere Dimers and Chromosome Looping in Human Spermatozoa. *Chromosome Res.* **2005**, *12*, 817–823. [CrossRef]

8. Ward, W.; Coffey, D. DNA Packaging and Organization in Mammalian Spermatozoa: Comparison with Somatic Cells. *Biol. Reprod.* **1991**, *44*, 569–574. [CrossRef]

9. Steger, K. Transcriptional and Translational Regulation of Gene Expression in Haploid Spermatids. *Anat. Embryol.* **1999**, *199*, 471–487. [CrossRef]

10. Carrell, D.T.; Emery, B.R.; Hammoud, S. Altered Protamine Expression and Diminished Spermatogenesis: What Is the Link? *Hum. Reprod. Update* **2007**, *13*, 313–327. [CrossRef]

11. De Vries, M.; Ramos, L.; Housein, Z.; De Boer, P. Chromatin Remodelling Initiation during Human Spermiogenesis. *Biol. Open* **2012**, *1*, 446–457. [CrossRef] [PubMed]

12. Van Der Heijden, G.W.; Dieker, J.W.; Derijck, A.A.H.A.; Muller, S.; Berden, J.H.M.; Braat, D.D.M.; Van Der Vlag, J.; De Boer, P. Asymmetry in Histone H3 Variants and Lysine Methylation between Paternal and Maternal Chromatin of the Early Mouse Zygote. *Mech. Dev.* **2005**, *122*, 1008–1022. [CrossRef] [PubMed]

13. Wright, W.E.; Piatyszek, M.A.; Rainey, W.E.; Byrd, W.; Shay, J.W. Telomerase Activity in Human Germline and Embryonic Tissues and Cells. *Dev. Genet.* **1996**, *18*, 173–179. [CrossRef]

14. Turner, S.; Hartshorne, G.M. Telomere Lengths in Human Pronuclei, Oocytes and Spermatozoa. *Mol. Hum. Reprod.* **2013**, *19*, 510–518. [CrossRef] [PubMed]

15. Baird, D.M.; Britt-Compton, B.; Rowson, J.; Amso, N.N.; Gregory, L.; Kipling, D. Telomere Instability in the Male Germline. *Hum. Mol. Genet.* **2006**, *15*, 45–51. [CrossRef] [PubMed]

16. Kimura, M.; Cherkas, L.F.; Kato, B.S.; Demissie, S.; Hjelmborg, J.B.; Brimacombe, M.; Cupples, A.; Hunkin, J.L.; Gardner, J.P.; Lu, X.; et al. Offspring's Leukocyte Telomere Length, Paternal Age, and Telomere Elongation in Sperm. *PLoS Genet.* **2008**. [CrossRef] [PubMed]

17. Kozik, A.; Bradbury, E.M.; Zalensky, A. Increased Telomere Size in Sperm Cells of Mammals with Long Terminal (TTAGGG)(n) Arrays. *Mol. Reprod. Dev.* **1998**, *51*, 98–104. [CrossRef]

18. Pickett, H.A.; Henson, J.D.; Au, A.Y.M.; Neumann, A.A.; Reddel, R.R. Normal Mammalian Cells Negatively Regulate Telomere Length by Telomere Trimming. *Hum. Mol. Genet.* **2011**, *20*, 4684–4692. [CrossRef] [PubMed]

19. Johnson, G.D.; Lalancette, C.; Linnemann, A.K.; Leduc, F.; Boissonneault, G.; Krawetz, S.A. The Sperm Nucleus: Chromatin, RNA, and the Nuclear Matrix. *Reproduction* **2011**, *141*, 21–36. [CrossRef]

20. Zalenskaya, I.A.; Bradbury, E.M.; Zalensky, A.O. Chromatin Structure of Telomere Domain in Human Sperm. *Biochem. Biophys. Res. Commun.* **2000**, *279*, 213–218. [CrossRef]

21. World Health Organization. *WHO Laboratory Manual for the Examination and Processing of Human Semen*, 5th ed.; World Health Organization: Geneva, Switzerland, 2010.

22. Zini, A.; Albert, O.; Robaire, B. Assessing Sperm Chromatin and DNA Damage: Clinical Importance and Development of Standards. *Andrology* **2014**, *2*, 322–325. [CrossRef] [PubMed]

23. Verhulst, S.; Susser, E.; Factor-litvak, P.R.; Simons, M.J.P.; Benetos, A.; Steenstrup, T.; Kark, J.D.; Aviv, A.; Lorraine, D.; Bank, D.; et al. Commentary: The Reliability of Telomere Length Measurements. *Int. J. Epidemiol.* **2015**, *44*, 4–7. [CrossRef] [PubMed]

24. Dan, F.; Eisenberg, T.A. Letters to the Editor Telomere Length Measurement Validity: The Coefficient of Variation Is Invalid and Cannot Be Used to Compare Quantitative Polymerase Chain Reaction and Southern Blot Telomere Length Measurement Techniques. *Int. J. Epidemiol.* **2016**, *45*, 1295–1298. [CrossRef]

25. Ribas-Maynou, J.; Amengual, M.J.; Garcia-Peiró, A.; Lafuente, R.; Alvarez, J.; Bosch-Rue, E.; Brassesco, C.; Brassesco, M.; Benet, J. Sperm Telomere Length in Motile Sperm Selection Techniques: A qFISH Approach. *Andrologia* **2017**, *50*, e12840. [CrossRef]

26. Boniewska-Bernacka, E.; Pańczyszyn, A.; Cybulska, N. Telomeres as a Molecular Marker of Male Infertility. *Hum. Fertil.* **2019**, *22*, 78–87. [CrossRef] [PubMed]

27. Ferlin, A.; Rampazzo, E.; Rocca, M.S.; Keppel, S.; Frigo, A.C.; De Rossi, A.; Foresta, C. In Young Men Sperm Telomere Length Is Related to Sperm Number and Parental Age. *Hum. Reprod.* **2013**, *28*, 3370–3376. [CrossRef] [PubMed]

28. Cariati, F.; Jaroudi, S.; Alfarawati, S.; Raberi, A.; Alviggi, C.; Pivonello, R.; Wells, D. Investigation of Sperm Telomere Length as a Potential Marker of Paternal Genome Integrity and Semen Quality. *Reprod. Biomed. Online* **2016**, *33*, 404–411. [CrossRef] [PubMed]

29. Thilagavathi, J.; Kumar, M.; Mishra, S.S.; Venkatesh, S.; Kumar, R.; Dada, R. Analysis of Sperm Telomere Length in Men with Idiopathic Infertility. *Arch. Gynecol. Obstet.* **2013**, *287*, 803–807. [CrossRef] [PubMed]

30. Singh, N.; Mishra, S.; Kumar, R.; Dada, R.; Malhotra, N. Mild Oxidative Stress Is Beneficial for Sperm Telomere Length Maintenance. *World J. Methodol.* **2016**, *6*, 163. [CrossRef]

31. Garolla, A.; Foresta, C.; Speltra, E.; Ferlin, A.; Rocca, M.S.; Menegazzo, M. Sperm Telomere Length as a Parameter of Sperm Quality in Normozoospermic Men. *Hum. Reprod.* **2016**, *31*, 1158–1163. [CrossRef]

32. Durairajanayagam, D. Lifestyle Causes of Male Infertility. *Arab J. Urol.* **2018**, *16*, 10–20. [CrossRef] [PubMed]

33. Liochev, S.I. Reactive Oxygen Species and the Free Radical Theory of Aging. *Free Radic. Biol. Med.* **2013**, *60*, 1–4. [CrossRef] [PubMed]

34. Aitken, R.J.; Smith, T.B.; Jobling, M.S.; Baker, M.A.; De Iuliis, G.N. Oxidative Stress and Male Reproductive Health. *Asian J. Androl.* **2014**, *16*, 31–38. [CrossRef] [PubMed]

35. Coluzzi, E.; Colamartino, M.; Cozzi, R.; Leone, S.; Meneghini, C.; O'Callaghan, N.; Sgura, A. Oxidative Stress Induces Persistent Telomeric DNA Damage Responsible for Nuclear Morphology Change in Mammalian Cells. *PLoS ONE* **2014**, *9*, e110963. [CrossRef] [PubMed]

36. Richter, T.; von Zglinicki, T. A Continuous Correlation between Oxidative Stress and Telomere Shortening in Fibroblasts. *Exp. Gerontol.* **2007**, *42*, 1039–1042. [CrossRef]

37. Moskovtsev, S.I.; Willis, J.; White, J.; Mullen, J.B.M. Disruption of Telomeretelomere Interactions Associated with DNA Damage in Human Spermatozoa. *Syst. Biol. Reprod. Med.* **2010**, *56*, 407–412. [CrossRef]

38. De Frutos, C.; Ló Pez-Cardona, A.P.; Balvís, N.F.; Laguna-Barraza, R.; Rizos, D.; Gutierrez-Adán, A.; Bermejo-Lvarez, P. Spermatozoa Telomeres Determine Telomere Length in Early Embryos and Offspring. *Reproduction* **2016**, *151*, 1–7. [CrossRef]

39. Jørgensen, P.B.; Fedder, J.; Koelvraa, S.; Graakjaer, J. Age-Dependence of Relative Telomere Length Profiles during Spermatogenesis in Man. *Maturitas* **2013**, *75*, 380–385. [CrossRef]

40. Wallenfang, M.R.; Nayak, R.; DiNardo, S. Dynamics of the Male Germline Stem Cell Population during Aging of Drosophila Melanogaster. *Aging Cell* **2006**, *5*, 297–304. [CrossRef]

41. Siderakis, M.; Tarsounas, M. Telomere Regulation and Function during Meiosis. *Chromosom. Res.* **2008**, *15*, 667–679. [CrossRef]

42. Nordfjäll, K.; Larefalk, A.; Lindgren, P.; Holmberg, D.; Roos, G. Telomere Length and Heredity: Indications of Paternal Inheritance. *Proc. Natl. Acad. Sci. USA* **2005**, *102*, 16374–16378. [CrossRef] [PubMed]

43. Broer, L.; Codd, V.; Nyholt, D.R.; Deelen, J.; Mangino, M.; Willemsen, G.; Albrecht, E.; Amin, N.; Beekman, M.; De Geus, E.J.C.; et al. Meta-Analysis of Telomere Length in 19 713 Subjects Reveals High Heritability, Stronger Maternal Inheritance and a Paternal Age Effect. *Eur. J. Hum. Genet.* **2013**, *21*, 1163–1168. [CrossRef] [PubMed]

44. Eisenberg, D.T.A.; Tackney, J.; Cawthon, R.M.; Theresa, C.; Kristen, C. Paternal and Grandpaternal Ages at Conception and Descendant Telomere Lengths in Chimpanzees and Humans. *Am. J. Phys. Anthropol.* **2017**, *162*, 201–207. [CrossRef] [PubMed]

45. Bauch, C.; Boonekamp, J.J.; Korsten, P.; Mulder, E.; Verhulst, S. Epigenetic Inheritance of Telomere Length in Wild Birds. *PLoS Genet.* **2019**, *15*, e1007827. [CrossRef] [PubMed]

46. Paul, C.; Robaire, B. Ageing of the Male Germ Line. *Nat. Rev. Urol.* **2013**, *10*, 227. [CrossRef]

47. Bryant, J.M.; Meyer-ficca, M.L.; Dang, V.M.; Berger, S.L.; Meyer, R.G. Separation of Spermatogenic Cell Types Using STA-PUT Velocity Sedimentation. *JoVE* **2013**, *80*, e50648. [CrossRef]

48. Callaghan, N.J.O.; Fenech, M. A Quantitative PCR Method for Measuring Absolute Telomere Length. *Biol. Proced. Online* **2011**, *13*, 3. [CrossRef]

49. Bebbington, K.; Spurgin, L.G.; Fairfield, E.A.; Dugdale, H.L.; Komdeur, J.; Burke, T.; Richardson, D.S. Telomere Length Reveals Cumulative Individual and Transgenerational Inbreeding Effects in a Passerine Bird. *Mol. Ecol.* **2016**, *25*, 2949–2960. [CrossRef]

50. Fujita, T.; Yuno, M.; Okuzaki, D.; Ohki, R.; Fujii, H. Identification of Non-Coding RNAs Associated with Telomeres Using a Combination of enChIP and RNA Sequencing. *PLoS ONE* **2015**, *10*, e0123387. [CrossRef]

51. Sullivan, R. Epididymosomes: A Heterogeneous Population of Microvesicles with Multiple Functions in Sperm Maturation and Storage. *Asian J. Androl.* **2015**, *17*, 726. [CrossRef]

52. Schoeftner, S.; Blasco, M.A. Chromatin Regulation and Non-Coding RNAs at Mammalian Telomeres. *Semin. Cell Dev. Biol.* **2010**, *21*, 186–193. [CrossRef] [PubMed]

53. Azzalin, C.M.; Reichenbach, P.; Khoriauli, L.; Giulotto, E.; Lingner, J. Telomeric Repeat-Containing RNA and RNA Surveillance Factors at Mammalian Chromosome Ends. *Science* **2007**, *318*, 798–802. [CrossRef] [PubMed]

54. Luke, B.; Panza, A.; Redon, S.; Iglesias, N.; Li, Z.; Lingner, J. The Rat1p 5′ to 3′ Exonuclease Degrades Telomeric Repeat-Containing RNA and Promotes Telomere Elongation in Saccharomyces Cerevisiae. *Mol. Cell* **2008**, *32*, 465–477. [CrossRef] [PubMed]

55. Savitsky, M.; Kwon, D.; Georgiev, P.; Kalmykova, A.; Gvozdev, V. Telomere Elongation Is under the Control of the RNAi-Based Mechanism in the Drosophila Germline. *Genes Dev.* **2006**, *20*, 345–354. [CrossRef] [PubMed]

56. Montero, J.J.; López De Silanes, I.; Granã, O.; Blasco, M.A. Telomeric RNAs Are Essential to Maintain Telomeres. *Nat. Commun.* **2016**, *7*, 12534. [CrossRef] [PubMed]

57. Montero, J.J.; López-Silanes, I.; Megías, D.; Fraga, M.; Castells-García, Á.; Blasco, M.A. TERRA Recruitment of Polycomb to Telomeres Is Essential for Histone Trymethylation Marks at Telomeric Heterochromatin. *Nat. Commun.* **2018**, *9*, 4–6. [CrossRef] [PubMed]

58. Reig-Viader, R.; Vila-cejudo, M.; Vitelli, V.; Busca, R.; Ruiz-herrera, A.; Valle, C.; Uab, C.; Spallanzani, B.L. Telomeric Repeat-Containing RNA (TERRA) and Telomerase Are Components of Telomeres During Mammalian Gametogenesis. *Biol. Reprod.* **2014**, *90*, 1–13. [CrossRef] [PubMed]

59. Reig-Viader, R.; Capilla, L.; Vila-cejudo, M.; Ruiz-herrera, A. Telomere Homeostasis Is Compromised in Spermatocytes from Patients with Idiopathic Infertility. *Fertil. Steril.* **2014**, *102*. [CrossRef]

MDPI

St. Alban-Anlage 66

4052 Basel

Switzerland

Tel. +41 61 683 77 34

Fax +41 61 302 89 18

www.mdpi.com

Genes Editorial Office

E-mail: genes@mdpi.com

www.mdpi.com/journal/genes